Agricultural Process Engineering

Agricultural Process Engineering

Ajay Mehta

RANDOM PUBLICATIONS
NEW DELHI (INDIA)

Agricultural Process Engineering

ISBN 978-93-5111-368-3

Published in 2014 in India by

Reprint 2020

RANDOM PUBLICATIONS

4376-A/4B, Gali Murari Lal, Ansari Road
New Delhi-110 002
Phone : +91-11-43580356, +91-11-23289044
e-mail: randomexports@gmail.com, sales@randompublications.com,
info@randompublications.com

Type Setting by: Friends Media, Delhi-110089
Printed at : Mehra Printers, Delhi-110092

Preface

The role of agricultural engineering is increasing with the dawning of a new century. Agriculture will have to supply not only food, but also other materials such as bio-fuels, organic feed stocks for secondary industries of destruction, and even medical ingredients. Furthermore, new agricultural technology is also expected to help reduce environmental destruction. Agricultural engineers apply their knowledge of engineering technology and science to agriculture and the efficient use of biological resources. Accordingly, they also are referred to as biological and agricultural engineers. They design agricultural machinery, equipment, sensors, processes, and structures, such as those used for crop storage. Some engineers specialize in areas such as power systems and machinery design, structural and environmental engineering, and food and bioprocess engineering. Agricultural engineers develop and install agricultural, horticultural and forestry machinery and processes. They also advise farmers, companies and government departments on countryside issues. These could range from crop diversity through to sustainable land use.

The development of civilization is closely related to the improvement in self-dependence of nation on its demands of food and other agricultural products which are the minimum basic requirements of the human society. In the matter of this kind of improvement, Agricultural Engineering plays a very great role in respect of further development of food productivity, agro-based industries, post-harvesting, food processing, Soil & Water Engineering, irrigation and Drainage etc. through the utilization of the software based latest technology. With the advancement of agriculture, the education in Agricultural Engineering has also been gradually developed and taken a definite shape through inclusion of theoretical as well as practical courses in multidirectional fields of Soil & Water Engineering irrigation and Drainage Engineering, Land and Water Resource Development and Management, Rural Engineering, Aqua-cultural Engineering,

Farm Development and Management, Marketing, Marketing and Sales, Computer Applications etc. The Agricultural Engineering education integrates Engineering and Agricultural Science knowledge and skill to develop technology and / or process to raise production and productivity of agriculture and other farm produce through efficient utilization of natural resources and conserving the same for future use. Broadly, the activities include efficient utilization of agricultural inputs through improved techniques of soil and water management (conservation), efficient implements and machinery ensuring precision, timeliness and reduced drudgery in farming operations (mechanization), and improving quality of farm produce (processing and value additions).

The Agricultural Engineering education is dynamic and addresses issues relevant to social and technological development of the country. The quality and quantum of agricultural inputs, their management techniques, and also quality of farm produce and methods of value additions would keep on changing with advancement of industrialization in general and economic upliftment of farmers / processors in particular. With the advancement of technology related to agricultural inputs like seed, fertilizers, water and pesticides coupled with the equipment and energy sources for their application, it has become essential to optimize these to increase production and conserve natural resources. The Agricultural Engineers today are required to provide technology not only for increasing crop production but also for reducing post harvest losses and value additions through processing of produce. Increasing export opportunity in agricultural and allied industrial sector, besides diversification has put additional demand on agricultural engineering education to meet the growing requirement of food feed and fibre through efficient management of costly inputs.

This book is fed with the information of this subject. This book will be very useful for a wide range of interested groups.

I would like to thank my team for standing beside me throughout my career and writing this book. My special thanks go to "Random Publications" who have published the book.

—Ajay Mehta

Contents

1

Introduction to Agro-Processing

Agriculture is the cornerstone of most developing countries' economies. Unfortunately, agriculture alone is no longer able to provide a reliable livelihood for the growing populations in these countries. Aternative or additional income generating opportunities are needed to support the millions of poor families who can no longer support their livelihoods from the land alone.

Agro processing could be defined as set of techno-economic activities carried out for conservation and handling of agricultural produce and to make it usable as food, feed, fibre, fuel or industrial raw material.

Hence, the scope of the agro-processing industry encompasses all operations from the stage of harvest till the material reaches the end users in the desired form, packaging, quantity, quality and price. Ancient Indian scriptures contain vivid account of the post harvest and processing practices for preservation and processing of agricultural produce for food and medicinal uses.

Inadequate attention to the agro-processing sector in the past put both the producer and the consumer at a disadvantage and it also hurt the economy of the Country.

Agroprocessing is related to turning primary agricultural products into other commodities for market and has the potential to provide those opportunities.

Practical Action's agroprocessing projects aim to increase income and access to food for the poor, by establishing small-scale, appropriate and sustainable processing businesses that are flexible, require little capital investment and can be carried out in the home without the need for sophisticated or expensive equipment.

The overall potential of agroprocessing is huge. It can reduce wastage, enhance food security, improve livelihoods for low-income groups and empower women. In sub-Saharan Africa, for instance, it is estimated that 60 per cent of the labour force finds part of its work in small scale food processing, and the majority of them are women.

However, there are a number of constraints that limit the further development of small-scale food processing in developing countries. Starting and developing a small business in a developing country is no easy task. Entrepreneurs face many challenges, especially with the uncertainty that exists over access to finance, advice and information, and reliable markets.

Agro-processing is now regarded as the sunrise sector of the Indian economy in view of its large potential for growth and likely socio economic impact specifically on employment and income generation. Some estimates suggest that in developed countries, up to 14 per cent of the total work force is engaged in agro-processing sector directly or indirectly. However, in India, only about 3 per cent of the work force finds employment in this sector revealing its underdeveloped state and vast untapped potential for employment. Properly developed, agro-processing sector can make India a major player at the global level for marketing and supply of processed food, feed and a wide range of other plant and animal products.

Cleaning and Grading

Seed cleaning removes the chaff from the seed, which not only allows the seeds to be sowed through a mechanical planter, but also allows the seed to be stored with less of a chance of fungal/bacterial infection. Cleaning the seed also works as a process to check for the purity of the collection.

High quality, pure seed is a critical ingredient in efficient seedling production.

The process is not simple. Seed is run through machines to remove the chaff and excess plant material surrounding them. The machine chosen to clean a specific seed is based primarily on seed size and type of chaff that has to be removed from the seed. As shown below, there are many different machines with size- and dimension-calibrated screens to accommodate the variety of native seeds that must be cleaned. Typically, the seed will be put through more than one machine to ensure that all chaff and plant material is removed. After the seed has been cleaned it is stored in a cotton or synthetic bag.

Size Reduction and Milling

A grinding mill is a unit operation designed to break a solid material into smaller pieces. There are many different types of grinding mills and many types of materials processed in them. Historically mills were powered by hand (mortar and pestle), working animal (horse mill), wind (windmill) or water (watermill). Today they are also powered by electricity.

The grinding of solid matters occurs under exposure of mechanical forces that trench the structure by overcoming of the interior bonding forces. After the grinding the state of the solid is changed: the grain size, the grain size disposition and the grain shape.

Milling also refers to the process of breaking down, separating, sizing, or classifying aggregate material. For instance rock crushing or grinding to produce uniform aggregate size for construction purposes, or separation of rock, soil or aggregate material for the purposes of structural fill or land reclamation activities. Aggregate milling processes are also used to remove or separate contamination or moisture from aggregate or soil and to produce "dry fills" prior to transport or structural filling.

Grinding may serve the following purposes in engineering:

- increase of the surface area of a solid
- manufacturing of a solid with a desired grain size
- pulping of resources

Grinding Laws

In spite of a great number of studies in the field of fracture schemes there is no formula known which connects the technical grinding work with grinding results. To calculate the needed grinding work against the grain size changing three half-empirical models are used. These can be related to the Hukki relationship between particle size and the energy required to break the particles. In stirred mills, the Hukki relationship does not apply and instead, experimentation has to be performed to determine any relationship.

- Kick for $d > 50$ mm

$$W_K = c_k(\ln(d_A) - \ln(d_E))$$

- Bond for 50 mm $> d >$ 0.05 mm

$$W_B = c_B\left(\frac{1}{\sqrt{d_E}} - \frac{1}{\sqrt{d_A}}\right)$$

- Von Rittinger for $d < 0.05$ mm

$$W_R = c_R\left(\frac{1}{d_E} - \frac{1}{d_A}\right)$$

with W as grinding work in kJ/kg, c as grinding coefficient, d_A as grain size of the source material and d_E as grain size of the ground material.

A reliable value for the grain sizes d_A and d_E is d_{80}. This value signifies that 80% (mass) of the solid matter has a smaller grain size. The Bond's grinding coefficient for different materials can be found in various literature. To calculate the KICK's and Rittinger's coefficients following formulas can be used

$$c_K = 1.151 c_B (d_{BU})^{-0.5}$$

$$c_R = 0.5 c_B (d_{BL})^{0.5}$$

with the limits of Bond's range: upper d_{BU} = 50 mm and lower d_{BL} = 0.05 mm.

To evaluate the grinding results the grain size disposition of the source material (1) and of the ground material (2) is needed. Grinding degree is the ratio of the sizes from the grain disposition. There are several definitions for this characteristic value:

- Grinding degree referring to grain size d_{80}

$$Z_d = \frac{d_{80,1}}{d_{80,2}}$$

Instead of the value of d_{80} also d_{50} or other grain diameter can be used.

- Grinding degree referring to specific surface

$$Z_S = \frac{S_{v,2}}{S_{v,1}} = \frac{S_{m,2}}{S_{m,1}}$$

The specific surface area referring to volume S_v and the specific surface area referring to mass S_m can be found out through experiments.

- Pretended grinding degree

$$Z_a = \frac{d_1}{a}$$

The discharge die gap a of the grinding machine is used for the ground solid matter in this formula.

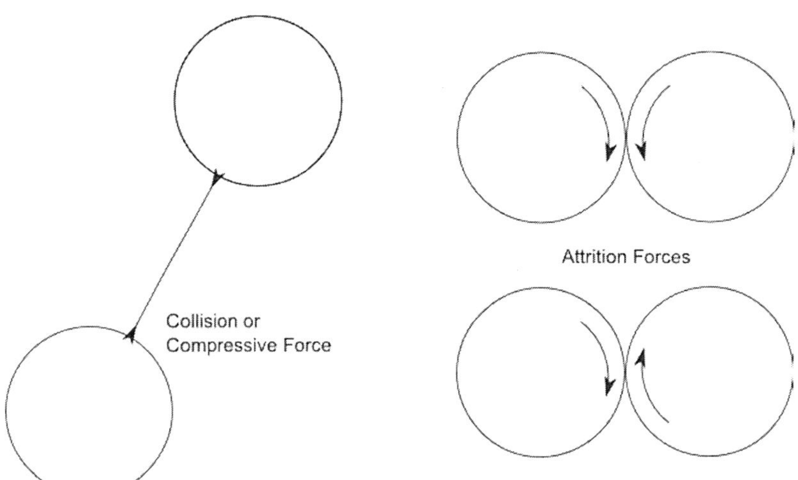

Grinding Machines

In materials processing a grinder is a machine for producing fine particle size reduction through attrition and compressive forces at the grain size level. In general, grinding processes require a relatively large amount of energy; for this reason, an experimental method to measure the energy used locally during milling with different machines was recently proposed.

Figure: *Operation of a ball mill*

Ball Mill

A typical type of fine grinder is the ball mill. A slightly inclined or horizontal rotating cylinder is partially filled with balls, usually stone or metal, which grinds material to the necessary fineness by friction and impact with the tumbling balls. Ball mills normally operate with an approximate ball charge of 30%. Ball mills are characterized by their smaller (comparatively) diameter and longer length, and often have a length 1.5 to 2.5 times the diameter. The feed is at one end of the cylinder and the discharge is at the other. Ball mills are commonly used in the manufacture of Portland cement and finer grinding stages of mineral processing. Industrial ball mills can be as large as 8.5 m (28 ft) in diameter with a 22 MW motor, drawing approximately 0.0011% of the total world's power. However, small versions of ball mills can be found in laboratories where they are used for grinding sample material for quality assurance.

The power predictions for ball mills typically use the following form of the Bond equation:

$$E = 10W\left(\frac{1}{\sqrt{P_{80}}} - \frac{1}{\sqrt{F_{80}}}\right)$$

where

- E is the energy (kilowatt-hours per metric or short ton)
- W is the work index measured in a laboratory ball mill (kilowatt-hours per metric or short ton)
- P_{80} is the mill circuit product size in micrometers
- F_{80} is the mill circuit feed size in micrometers.

Rod Mill

A rotating drum causes friction and attrition between steel rods and ore particles. But note that the term 'rod mill' is also used as a synonym for a slitting mill, which makes rods of iron or other metal. Rod mills are less common than ball mills for grinding minerals.

The rods used in the mill, usually a high-carbon steel, can vary in both the length and the diameter. However, the smaller the rods, the larger is the total surface area and hence, the greater the grinding efficiency

Autogenous Mill

Autogenous mills are so-called due to the self-grinding of the ore: a rotating drum throws larger rocks of ore in a cascading motion which causes impact breakage of larger rocks and compressive grinding of finer particles. It is similar in operation to a SAG mill as described below but does not use steel balls in the mill. Also known as ROM or "Run of Mine" grinding.

SAG Mill

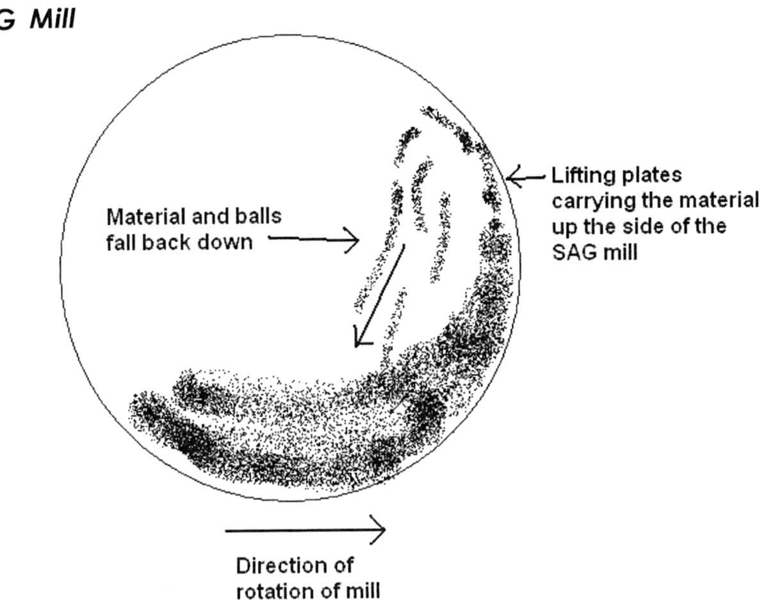

Figure: *Principle of SAG Mill operation*

SAG is an acronym for Semi-Autogenous Grinding. SAG mills are essentially autogenous mills, but utilize grinding balls to aid in grinding like in a ball mill. A SAG mill is generally used as a primary or first stage grinding solution. SAG mills use a ball charge of 8 to 21%. The largest SAG mill is 42' in diameter, powered by a 28 MW (38,000 HP) motor. A SAG mill with a diameter 44' in diameter has been designed with a power of 35 MW (47,000 HP).

Attrition between grinding balls and ore particles causes grinding of finer particles. SAG mills are characterized by their large diameter and short length as compared to ball mills. The inside of the mill is lined with lifting plates to lift the material inside the mill, where it then falls off the plates onto the rest of the ore charge. SAG mills are primarily used at gold, copper and platinum mines with applications also in the lead, zinc, silver, alumina and nickel industries.

Pebble Mill

A rotating drum causes friction and attrition between rock pebbles and ore particles. May be used where product contamination by iron from steel balls must be avoided. Quartz or silica is commonly used because it is inexpensive to obtain.

High Pressure Grinding Rolls

The high pressure grinding rolls, often referred to as HPGRs or roller press, consists out of two rollers with the same dimensions, which are rotating against each other with the same circumferential speed. The special feeding of bulk material through a hopper leads to a material bed between the two rollers. The bearing units of one roller can move linearly and are pressed against the material bed by springs or hydraulic cylinders. The pressures in the material bed are greater than 50 MPa. In general they achieve 100 to 300 MPa. By this the material bed is compacted to a solid volume portion of more than 80%.

The roller press has a certain similarity to roller crushers and roller presses for the compacting of powders, but purpose, construction and operation mode are different.

Extreme pressure causes the particles inside of the compacted material bed to fracture into finer particles and also causes microfracturing at the grain size level. Compared to ball mills HPGRs are achieving a 30 to 50% lower specific energy consumption, although they are not as common as ball mills since they are a newer technology.

A similar type of intermediate crusher is the edge runner, which consists of a circular pan with two or more heavy wheels known as mullers rotating within it; material to be crushed is shoved underneath the wheels using attached plow blades.

Buhrstone Mill

Another type of fine grinder commonly used is the buhrstone mill, which is similar to old-fashioned flour mills.

Vertical shaft impactor mill (VSI mill)

Type of fine grinder which uses a free impact of rock or ore particles with a wear plate. High speed of the motion of particles is achieved with a rotating accelerator. This type of mill uses the same principle as VSI Crusher

Tower Mill

Tower mills, often called vertical mills, stirred mills or regrind mills, are a more efficient means of grinding material at smaller particle sizes, and can be used after ball mills in a grinding process. Like ball mills, grinding (steel) balls or pebbles are often added to stirred mills to help grind ore, however these mills contain a large screw mounted vertically to lift and grind material. In tower mills, there is no cascading action as in standard grinding mills. Stirred mills are also common for mixing quicklime (CaO) into a lime slurry. There are several advantages to the tower mill: low noise, efficient energy usage, and low operating costs.

Grinding

Grinding or particle-size reduction is a major function of feed manufacturing. Many feed mills pass all incoming ingredients through a grinder for several reasons:

(a) clumps and large fragments are reduced in size,

(b) some moisture is removed due to aeration, and

(c) additives such as antioxidants may be blended.

All of these improve the ease of handling ingredients and their storability.

There are other reasons for grinding and the associated sieving of ingredients in formula feeds before further processing. Small fish and fry require plankton-size feeds available in dry form as a meal or granule. Extremes in particle sizes are wasteful and often dangerous. Fry have been killed because of their inability to pass through the digestive system large pieces of connective tissue and bone present in dry animal byproducts, or hull fragments found in cottonseed meal and rice bran. On the other hand, dust or "fines" may become colloidal suspensions in water, so dilute that several mouthfuls carry little nutritive value.

The grinding of ingredients generally improves feed digestibility, acceptability, mixing properties, pelletability, and increases the bulk density of some ingredients. It is accomplished by many types of manual and mechanical operations involving impact, attrition, and cutting.

Hammermills

Hammermills are mostly impact grinders with swinging or stationary steel bars forcing ingredients against a circular screen or solid serrated section designated as a striking plate. Material is held in the grinding chamber until it is reduced to the size of the openings in the screen. The number of hammers on a rotating shaft, their size, arrangement, sharpness, the speed of rotation, wear patterns, and clearance at the tip relative to the screen or striking plate are important variables in grinding capacity and the appearance of the product. Heat imparted to the material, due to the work of grinding, is related to the time it is held within the chamber and the air flow characteristics. Impact grinding is most efficient with dry, low-fat ingredients, although many other materials may be reduced in size by proper screen selection and regulated intake.

Most hammermills have a horizontal drive shaft which suspends vertical hammers but for some ingredients, such as dried animal byproducts, a "vertical" hammermill is more efficient. In this mill, the drive shaft is positioned vertically and screens and hammers are positioned horizontally. Material successfully reduced in size to the diameter of screen holes or smaller, are carried by gravity outside the mill and thence by air or conveyor to storage in "make-up" bins. Oversize particles, not easily broken, drop through the mill and may be re-cycled or discarded. Thus foreign materials, such as metal and stones, are discharged before they are forced through the screen causing damage.

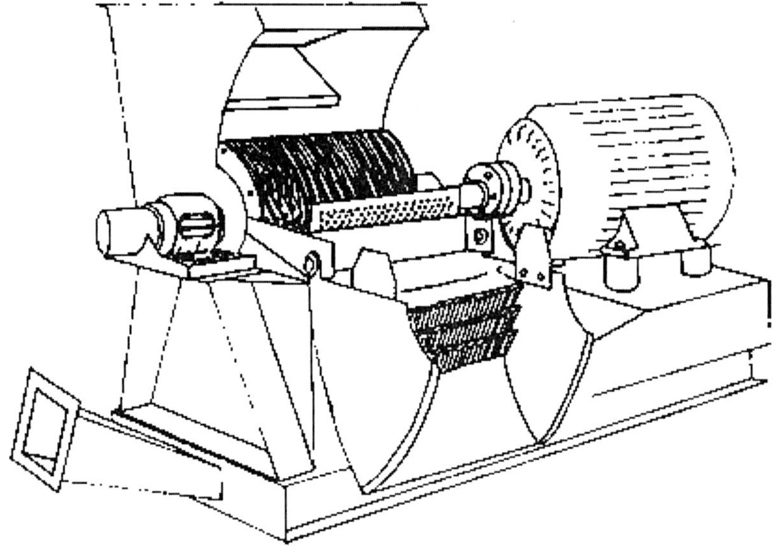

Figure: *Hammer Mill*

Attrition Mills

Attrition mills use the hammermill principle to a certain extent; i.e., shattering by/impact. However, they also impart a shearing and cutting action. Grinding is done between two discs equipped with replaceable wearing surfaces. One or both of these discs is rotated; if both, they rotate in opposite directions. When one disc is rotated, and the other stationary, the assembly is used for shredding and deferring. Often materials which have been coarsely ground by other mills, are passed through an attrition mill for blending or smoothing out an ingredient or mixture containing liquids which may have clumps. The discs of an attrition mill are generally in a vertical position so that materials not capable of reduction can pass by gravity out of the grinding area.

Roller Mills

A combination of cutting, attrition, and crushing occurs in roller mills. These are smooth or corrugated rolls rotating at the same speed set at a pre-determined distance apart with material passing between the two. A tearing action may be added by operating the rolls at different speeds and by corrugations which are different for each roll; i.e., the top roll may have off-radial spiral corrugations and the bottom roll lateral corrugations. This last type, called a "Le Page cut" is used in making granules from hard pellets, as it provides a breaking surface without much impact to cause dust. Roll grinding is economical but limited to materials which are fairly dry and low in fat.

Cutters

Rotary cutters are a type of grinder which reduces dry particle solids mainly by shearing with knife edges against a striking plate. The mill also includes the processes of attrition and impact, although these actions are limited if the material is easily reduced by cutting and the screen limiting discharge has large perforations. The mill consists of a rotating shaft with four attached parallel knives and a screen occupying one fourth of the 360 degree rotation. The mill is best used to crack whole grains with a minimum of "fines". It is not used as a final process for reducing the size of ingredients used in fish feeds.

Screening

Associated with grinding feeds for fish fry, a sieving system is required which classifies materials to any desired particle size. The "overs" in this system may be re-ground or rejected. The "throughs" may be selected to comply with fish preferences for size and mixed according to formula specifications. Feeds sifted through a 177-micron opening (a U.S. No. 100 sieve) have been successfully used for increasing survival and growth of minnows and catfish fry. Hammermill or impact grinding of dry feeds, especially cereal grains, creates particles within the range called "dust", and a dust-collecting system may be necessary to remove this. An excess of dust in the feed may lead to gill disease, a situation where organic matter adhering to gills becomes a nutrient for bacteria or parasites.

The problem of excess dust formed by grinding feeds may be partly alleviated by adding a spray of oil or a semi-moist ingredient, such as condensed fish solubles or fermentation solubles, on feeds entering the grinder. Dehydrated alfalfa is prepared as a dust-free meal, similar in texture to a sifted crumblized pellet, by spraying mineral oil into a hammermill chamber during grinding.

Mixing

The objective of feed mixing is to start with a certain assortment of ingredients called a "formula", totalling some definite weight. This is processed so that each small unit of the whole, either a mouthful or a day's feeding, is the same proportion as the original formula. Mixing is recognized as an empirical unit operation, which means that it is more of an art than a science and must be learned by experience.

Feed mixing may include all possible combinations of solids and liquids. Within each ingredient are differences in physical properties. For solids there are differences in particle size, shape, density, electostatic charge, coefficient of friction as represented by the angle of repose, elasticity or resilience and, of course, colour, odour, and taste. For liquids there are differences in viscosity and density.

The term "mixed" can mean either blended, implying uniformity, or made up of dissimilar parts, implying scattering. As applied to formula feeds, the objective of mixing combines each of these definitions; i.e., the scattering of dissimilar parts into a blend. However, it is improbable that uniformity is attained with particles within a, sample arranged in some order of position or concentration. That is only a quality control; goal. It has been suggested that a proper title for a discussion of mixing should be "mixing and unmixing", for during the operation there is a constant tendency of particles which have been mixed to become separated. Three mechanisms are involved in the mixing process:

(a) the transfer of groups of adjacent particles from one location in the mass to another,

(b) diffusion: distribution of particles over a freshly developed surface,

(c) shear: slipping of particles between others in the mass.

In theory, the position of particles within a container is determined by chance, and the effects of chance accumulate until they outweigh the direct effects of mixing action. In the mixing of liquids, chance movement of components creates order or uniformity. With dry solids, chance distribution creates disorder. When disorder is at a more or less stable maximum, it may be called "random". Many factors in dry solids cause particles to avoid a chance or random arrangement. In fact, the result of mixing feed ingredients may be a definite pattern of particle segregation or non-random arrangement.

Particle segregation is due to differences in the physical properties of ingredients and the shape and surface characteristics of the mixer.

Particle size may be the most important factor in causing segregation. An improvement in mixing which approaches random distribution of solids by decreasing particle size can be measured quantitatively by statistical methods. In general, the smaller and the more uniformally sized the ingredients are prepared, the more nearly they will approach random distribution during mixing.

In many formulae, a decrease in particle size is necessary to attain a sufficient number of particles of an essential additive (vitamin, mineral, medication) for dispersion in each daily feed unit. This may require the particle size to be the diameter of dust, 10 to 50 microns. Certain ingredients are unstable in finely divided form and likely to acquire an electrostatic charge. Concentration of particles on a charged surface, roughness of the mixed and stickiness of oily and wet ingredients are factors in causing segregation when very small particles are mixed and when these are much smaller than the bulk of other ingredients.

Mixing may be either a batch or a continuous process. Batch mixing can be done on an open flat surface with shovels or in containers shaped as cylinders, half-cylinders, cones or twin-cones with fixed baffles or moving augers, spirals, or paddles. Continuous mixing proportions by weight or volume, is a technique best suited for formula feeds with few ingredients and minimal changes.

Horizontal Mixers

Continuous Ribbon Mixers

The continuous or "twin-spiral" mixer consists of a horizontal, stationary, half-cylinder with revolving helical ribbons placed on a central shaft so as to move materials from one end to the other as the shaft and ribbon rotate inside. Capacity can be from a few litres to several cubic metres. The speed of shaft rotation will vary inversely as the circumference of the outer ribbon; usually optimum between 75-100 metres per minute. Since material travel is from one end to the other, either end may be used for discharge. These mixers may be inverted for cleaning.

Non-continuous Ribbon Mixers

Non-continuous or interrupted ribbons are similar to the continuous ribbon mixers except that short sections called "paddles" or "ploughs" are spaced in a spiral round the mixer shaft. Action is different from that of continuous ribbon mixers, and may be more satisfactory for mixing liquids with dry solids. These mixers are made in a wide variety of sizes with travel of the outer diameter of paddles from 100 to 120 metres per minute.

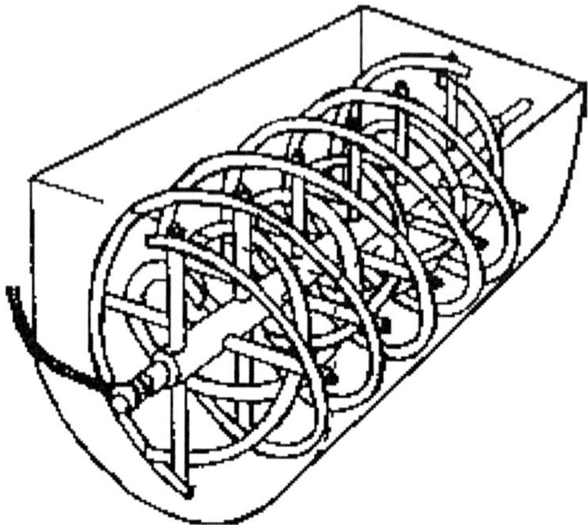

Figure: *Continuous ribbon mixer*

Vertical Mixers

Vertical mixers may consist of a cylinder, cone, or hopper-shaped container, with a single or double screw (auger) located vertically through the centre. The screw operates at speeds of 100 to 200 rpm and vertically conveys incoming materials from the bottom (generally the intake) end, like a screw conveyor, to the top where they are scattered and fall by gravity. This sequence is repeated several times until a blend is attained (usually from 10 to 12 minutes). These mixers may also be loaded from the top. Results show that vertical mixers are not efficient for uniform mixing of solids and liquids or for materials of quite different particle size or density. This unit is difficult to clean and there may be inter-batch contamination.

Other Types of Mixers

A third type of mixer is the horizontal revolving drum. This can be a straight-sided cylinder or a cylinder tapered at each end. The sides may be smooth or fixed with baffles or shelves to pick up and drop ingredients. Smooth, dry materials of uniform physical properties are blended best in this type of mixer.

A modification of this type is the turbine mixer which is a fixed cylinder with revolving shaft to which are fixed paddles, ploughs, scrapers, or shelves designed to re-pile materials continually. This mixer is often used as a cooker to dry fish wastes and to blend various types of fish meal into a standardized product. They are also particularly efficient for mixing heavy ingredients and for adding liquids to mixtures which would clump or cake in another type of mixer. Some particle size reduction (grinding) may occur on soft materials, such as rice bran and alfalfa leaves. A complete mixing can usually be attained in 3 to 6 minutes unless longer time is necessary to eliminate lumps caused by added liquids. Mixer shaft rotation is regulated to provide some centrifugal action, but this must not be excessive.

The "Nauta" mixer originated in Holland and is constructed in the form of an inverted cone with a mixing screw inside rotating around the inside wall. The mixer is made in a variety of sizes from laboratory models, for premixing chemical and vitamin additives, to very large production sizes. It is excellent for premixing trace elements and works very well for adding moderate amounts of liquids into dry ingredients.

Another type of mixer called the "entoleter" consists of a high-speed rotating disc which throws the ingredient charge with considerable force against the walls of a chamber. This mixer functions well to smooth out clumps or balls of compacted ingredients and will cause eggs of grain weevils to become inactive. However, since it may shatter vitamin A beadlets encapsulated in gelatine, it is not recommended for all mixtures.

Liquid Mixers

Oils and water-miscible oil preparations are often added to dry ingredients as a source of energy or as a specific nutrient. Although the oil-soluble vitamins. A, D, E, and K, are available in dry carrier concentrates, they may be obtained in pure form and premixed by the feed manufacturer. Liquids containing nutrients can be mixed faster and with more uniformity than the same nutrient in dry concentrate condition. Therefore, a liquid blender may be needed in the feed plant.

Liquid blenders usually consist of a horizontal tub or cylinder with a number of wires or paddles equally spaced around a shaft which revolves inside. Sometimes the shaft is hollow and liquids are forced through holes in the paddles in a spray effect. Some models have a shaft speed of 400 to 600 rpm while others rotate at 1 200 rpm. Ingredients such as condensed fish or fermentation solubles, molasses, or fish oils are often premixed in a bowl type variable speed mixer, blending the liquid with dry ingredients.

Mixing Operation and Evaluation

Accurate mixing requires the addition of ingredients in a tested sequence from batch to batch. The usual practice is to add large-volume ingredients first, then those of smaller amount. Unless already premixed, liquids should be added after all dry ingredients have been mixed. Total mixing time is critical and is influenced by the composition of the formula. All mixers should be calibrated by laboratory recovery of known additives (physically or chemically) so that under and over mixing does not occur. Uniformly sized salt, graphite, or iron particles coated with water soluble dyes are often used as "tracers". Each mixer should be calibrated for its mixing time and capacity by volume for best results.

Pelleting

The transformation of a soft, often dusty feed into a hard pellet is accomplished by compression, extrusion, and adhesion. The general process involves passing a feed mixture through a conditioning chamber where 4 to 6 percent water (usually as steam) may be added. Moisture provides lubrication for compression and extrusion and in the presence of heat causes some gelatinization of raw starch present on the surface of vegetative ingredients, resulting in adhesion. Within 20 seconds of entering the pellet mill, feed goes from an air-dry (about 10-12 percent moisture) condition at ambient temperature, to 15-16 percent moisture at 80-90°C. During subsequent compression and extrusion through holes in a ring' die, friction further increases feed temperature to nearly 92°C. Pellets discharged onto a screen belt of a horizontal tunnel drier or into a vertical screened hopper are air-cooled within 10 minutes to slightly above ambient temperatures and dried to below 13 percent moisture.

Contrary to early belief, finished pellets contain practically all the nutrients found in feedstuffs and additives as compounded. The loss of thermolabile vitamins used in additives, which may be slight or extensive in the case of vitamin C, may be compensated for by extra supplementation of these in the vitamin premix to comply with formula requirements. Diastatic enzymes (alpha and beta amylase) present in whole grains and cereal byproducts are still active after processing by grinding and pelleting, although powdered enzymes added as an ingredient are inactivated.

Application

Mechanically, the process of pelleting involves forcing soft feed through holes in a metal ring-type die. These holes may be round or square, tapered or non-tapered. Single or double rolls mounted inside the die ring on a cam or eccentric, turn on a rotating shaft as friction develops (due to the presence of feed between roll and die). Feed is forced through the die holes in increments so that dissection of a finished pellet shows tight layers of feed mixture. The die is driven by a motor and the rolls turn only as feed between rolls and die develops friction.

To make dry feed particles pliable for close compression and to decrease friction and absorb mechanical heat, water is often added to the feed, either as the formula is mixed or in the conditioning chamber of the pellet mill. Water may be provided as liquid and/or vapour. If water is provided in the form of steam, two objectives are accomplished:

(a) a high volume of water vapour condenses on the surface of feed particles, wetting and softening them for better compression, and

(b) the high temperatures of this steam cause some gelatinization of raw starch present in all vegetative ingredients, providing the necessary adhesion for firm pellets.

If sufficient moisture cannot be added as steam, pretreatment with water may be used to gain the desired lubricating effect.

It is not absolutely necessary to add steam to a soft feed in order to compress it into a pellet. Materials such as rice bran, ground cottonseeds, and palm kernel cake may be pelleted with no added moisture. The resulting pellets are often slightly charred from high temperature and the electrical energy consumption is much more than would be needed if moisture were added. The high fat content of these materials provides lubrication, but this does not soften the fibre sufficiently to avoid excessive heating caused by friction. The ring die and rolls of a pellet mill exposed to high temperatures show metal fatigue and must be replaced often. The proper conditioning of dehydrated alfalfa meal will permit the manufacture of over 2 000 tons of pellets during the life of a standard die of 40 mm thickness.

Overall, the texture of a soft feed mixture is changed from a meal-like material with bulk density approximately 0.4 g/cc, to a pellet with bulk density of 0.5 - 0.6 g/cc. Within the ring die, pressures of 75-600 kg/cm^2 are attained. Feed mixtures containing large amounts of fibrous ingredients often result in pellets too hard for gastric breakdown and digestion in fish. On the other hand, high-fat feeds and an excess of moisture cause poor quality pellets. Pellet quality may be defined as a certain hardness or water stability which assures efficient use without loss in handling on land or in water.

Feed formulation and operation of the pellet mill may be balanced to supply fish with a feed that is acceptable, available, and easily digested. The inter-dependent variables present in ingredient selection are subjects for research in each area of fish culture and for each fish species. Variables resulting from mechanical operation of the equipment may be listed here:

(a) Pellet die thickness as related to diameter of hole is a factor in pellet quality.

(b) Speed of rotation should also be considered for each die thickness/hole diameter combination.

(c) The speed at which feed is introduced into the feed-conditioning chamber affects the moisture/temperature relationship which in turn relates to pellet quality.

(d) The amount of steam added to a given volume of feed should be in balance such that the drive motor on the pellet mill is operating at its maximum amperage. The opening of the steam valve at the pellet mill has a direct relation to the amount of water entering the feed as steam at any given steam pressure.

(e) Atmospheric conditions in the factory, especially relative humidity, which pre-condition the ingredients before processing, will affect die selection and operational settings.

All of the above items must be examined for each feed formula, which shows that pelleting is more of an art than a science. It should be emphasized again that the pellet mill operates most efficiently when the motor amperage use is optimum for the voltage available. It is important to watch the ammeter gauge frequently during pelleting. At the start of a pelleting operation, a small amount of soft dry feed enters the pellet chamber for compression and the gauge will respond to the load with less than optimum results. The addition of steam results in a lowered reading, showing the lubricating effect. More dry feed may now enter the conditioning chamber and, as it reaches the die, the ammeter will move to a higher reading. Additional steam will lower this. By adjusting the dry feed intake and the steam valve opening, a stable condition will result where the motor is operating at its maximum rated amperage and pellet production will be at maximum capacity. These operational conditions normally coincide with maximum pellet quality afforded by the composition of the feed mix.

Although power transmission by means of a V-belt drive is shown (1), gear drive coupling of motor to the main shaft (2) is also possible. Feed from a surge bin (3) is metered into the steam conditioning chamber by a variable speed screw feeder (4). Paddles shown in the steam conditioning chamber (5) agitate the feed to ensure even blending of feed and steam. The conditioned feed is then fed by means of a distributor auger (6) into the pelletizing chamber (7) where it is extruded through the die (8). As the die rotates, feed is pressed against its inner wall by a set of 3 rollers (9). Due to its high rotational speed and heavy load, the die has to be securely harnessed to the shaft. This is achieved by means of a two-piece die cover assembly (10) and 12 strong bolts (11). The rollers usually have a hardened shell with indentations on the surface to provide traction and to reduce slippage of feed. Feed within the pelletizing chamber is continuously redistributed by adjustable feed ploughs (12). Extruded pellets of appropriate lengths are cut off by an assembly of knives (not shown) mounted on the inside of the die casing.

A control valve introduces dry steam into a header from which, through several port-holes, steam enters the conditioning chamber in contact with dry feed. Between this valve and the steam generator or boiler are a strainer and trap to remove condensate, providing only dry steam in the mill. At the discharge end of the conditioning chamber is a gate to restrict feed from immediately leaving and allows more time for moisture to be absorbed into the feed. A chute or funnel usually guides moisture-conditioned feed into the pellet chamber where compression and extrusion occur. If the ammeter goes much above the optimum reading, this chute may be quickly raised to prevent a choke-up of feed in the die holes. From the pellet chamber, feed may be directed by means of a butterfly valve to the cooler or on the floor for inspection.

Another variable that may be introduced into the operation of a pellet mill is the rotational speed of the die. For the production of small diameter pellets (i.e., 3 mm or less) high rotation speeds are used. This results in a thinner layer of soft feed inside the die ring ahead of the rolls, and for a given volume of feed the efficiency of pelleting and pellet hardness are improved. Die speeds may be changed by replacing the pulley on the main motor shaft of the pellet mill. Speeds generally range from 130-400 rpm. Feeds-of low bulk density are formed best in dies rotating at higher speeds.

Influence of Feed Composition

Additional factors known to influence pellet quality include bulk density of the soft feed, its texture, chemical composition (fat, fibre, carbohydrate, protein, and moisture) and prevailing ambient conditions of temperature and relative humidity. Grinding improves pellet quality by reducing air spaces between particles, allowing closer surface to surface contact for a given volume of feed; i.e., it increases bulk density. Large pieces of any ingredient in a feed formula result in weak spots in the pellet, especially if these are fibrous or bony. Grinding also increases the total surface area of a given weight of feed, thereby allowing more space for steam condensation during the conditioning process. This results in a higher feed temperature and more water absorption which together, within the time available, increases gelatinization of raw starch. Starch gelatinization has been mentioned previously, but since it is the key to water-stable pellets, further discussion may be beneficial. Starch gelatinization is the rupture of starch granules, thereby allowing the linear and cyclic molecules to hydrate and become sticky in the presence of water. Gelatinization occurs by mechanical means such as grinding, pressure, and by hot water. Soft feed at an environmental temperature of 25°C can be brought to a temperature of 85°C by the addition of 4 to 6 percent moisture from steam. Frictional heat due to passage of feed through the pellet mill adds 2 or 3 degrees of temperature. The moisture, temperature, and time involved combine to impart a sticky surface to starch-containing ingredients which, when subsequently dried, improved pellet hardness and water stability.

Since fat and moisture lubricate the soft feed as it passes through holes in the ring die, reducing friction of compression and extrusion, a certain level of these are necessary for pelleting. If the total crude fat in a feed formula is much more than 8-10 percent or the moisture is much above 15 percent, too much lubrication is provided and a poor quality pellet in terms of hardness results. Dies may be designed to accommodate high-fat feeds by tapering or slanting the Holes or by increasing the hole depth in relation to its diameter. Dietary fat in the final feed may be increased beyond that in the formula by spraying liquid fat or oil onto pellets or crumbles after they leave the pellet mill. An external coating of 5-6 percent stabilized fat does not appreciably soften the pellet except externally, and this may be a lure both in odour and texture for certain fishes.

In some countries, high humidity may cause problems in pelleting. Sun-dried ingredients and those which absorb moisture on storage (i.e. marine products contain sodium chloride which is hygroscopic) may need to be balanced with low-fat, fibrous ingredients or limited in their use. Low-fat cereals or cereal by-product cassava cake and pressed copra may be used in high-moisture, high-fat feeds to provide resistence to compression and extrusion (necessary to form a hard pellet). If the formula contains sugar or molasses, low-frictional heat is an asset, for carmelization occurs at about 60°C.

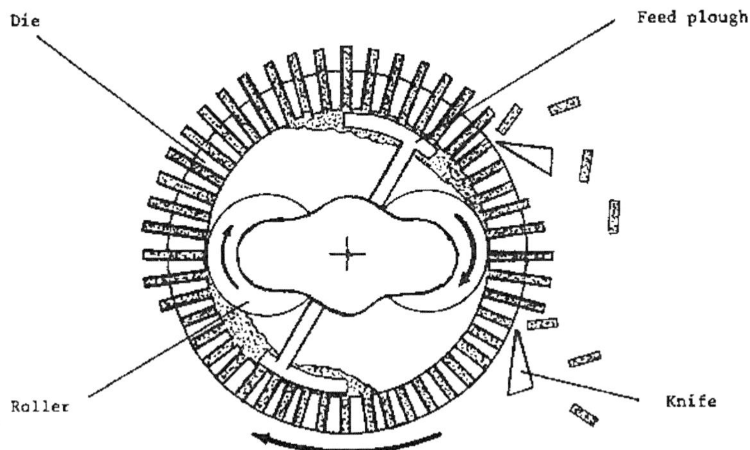

Figure: *Pellet extrusion (arrows indicate direction of rotation of moving parts)*

Cooling and Drying

The temperature imparted to pellets in the process of their manufacture assists the removal of moisture by the air-drying process. Generally, within ten minutes after extrusion, hard pellets are cooled to ambient temperature and brought to a moisture content slightly above that of the entering soft feed. This may be done by spreading pellets in a thin layer on the floor and blowing air over them. Commercially, it is done by passing the hot pellets through a vertical or a horizontal chamber designed to bring air at ambient temperature into intimate contact with the outer surface of the pellets.

Vertical cooler-dryer

Moisture added as steam, provides a large volume of lubricating material with a small weight. In addition, the temperature imparted to pellets by steam is a major factor in their subsequent drying. If air at ambient temperature is drawn or blown through the newly formed pellets, its ability to take up water depends on an increase in temperature. Air has an avidity for moisture directly related to its temperature. Air passing over hot pellets will increase in temperature and thus increase its water-holding ability. If cold or warm water were used to lubricate soft feed before pelleting, the final pellet temperature would not reach that necessary to gelatinize raw starch or significantly increase the ability of air to remove moisture in the cooler. Heated air may be used in the cooling-drying process, but this is generally uneconomical. Heating newly-formed, wet pellets on trays in an oven to attain drying does, however, result in very water-stable feed.

Pellets are discharged from the mill into the top of a flat-sided hopper and dropped into an attached cooling bin. This is divided in the middle with a plenum connected to the suction side of a blower fan. The weight of pellets filling the cooler pivots perforated louvers on the two sides to allow cool air to permeate the hot pellets, removing moisture, and cooling the pellets before entering the plenum for discharge through the blower. Pellets leave the bin at the bottom via discharge gates at a rate regulated by the amount of hot pellets entering the cooler. This ensures uniform cooling and drying of pellets.

Pellets gradually fill the cooler section of this system until they reach a diaphragm located near the top of the enclosed hopper. Pressure on this activates the discharge gates and the pellets are discharged until the level is below the diaphragm. A bucket elevator or drag screw conveys cooled pellets to either a storage or bagging bin or to a truck for bulk delivery. This type of cooler is preferred when space is limited and is generally more economical than the horizontal type.

Horizontal cooler-dryer

Coolers of the horizontal type consist of a moving wire belt or sectional belt of perforated metal trays which convey pellets from the discharge spout of the pellet mill . The depth of pellets on this belt and its speed of travel may be adjusted so that pellets leave for storage at a desired moisture and temperature. Horizontal coolers may be a single deck with pellets discharged at the end opposite the intake, or a double deck with two belts in the same enclosure; pellets return to the same end as they entered. Air from a centrifugal fan is made to flow from the cooler bottom through the layer of pellets. As in the vertical cooler, air is discharged into a dust-collecting system which removes the "fines" or particles which separate from the pellets. These are returned to the mill continuously for re-pelleting.

Crumbles

Cooled pellets may be ground on corrugated rolls and the resulting product sifted into various sizes of granules or crumbles. For small fish, the physical properties of crumbles are often more desired than a meal ration and easier to manufacture than a small pellet. Crumbles provide a multi-faceted surface to reflect light, this being a lure for sight feeders. They ensure that all the ingredients of a formula will be ingested, whereas components of meal feeds separate on entering water, allowing selection of certain ingredients, and the-solution or colloidal suspension of others.

Production of pellets of 4 mm or more in diameter proceeds at a higher rate than when feed is compressed into a smaller diameter. Pellet dies of small diameter holes are expensive and the time consumed in changing dies can make the manufacture of small amounts of fingerling feed quite costly. Generally, crumblizing rolls should be complementary equipment in any pelleting operation.

Screening or Grading

Some sifting is necessary in the production of pellets and crumbles. Small fragments (fines) are produced as hot, moist pellets are cut off from the die inside the pellet chamber, and as pellets pass through the cooling and conveying equipment. Fines may be returned to the pellet mill for reprocessing or used as feed for fry.

There are many types of sifting and grading systems both manual and mechanized. Most shake or rotate from side to side with material passing over screens of specified openings and covered to confine dust to the equipment. Sifting is the last process in manufacture of pellets and crumbles and this equipment should be located just above the bagging or final discharge bin.

Use of Hard Pellets

Test feeding in dirt-bottomed ponds or on feeding platforms with minnows, goldfish, catfish, trout, carp, and buffalo fish has shown that for fingerlings and larger fish, pellets provide a physical property which improves the economy of using artificial feeds. Very small fish have shown better growth and feed conversion with pelleted feeds, even those which disintegrate before ingestion, than with the same formula in meal form. The concentration of feed in a limited space and the inability of fish to select and reject certain ingredients, are factors in favour of using pellets. As fish grow larger, many species stop accepting small feed particles and suspended solids as supplied in meal-type feeds, and most of these “fines” are lost or become fertilizer. To be effective for some fish, feeds must retain a bite-size texture.

Drying and dehydration

Drying is a method of food preservation that works by removing water from the food, which inhibits the growth of bacteria and has been practiced worldwide since ancient times to preserve food. Where or when dehydration as a food preservation technique was invented has been lost to time, however the earliest known practice of food drying is 12,000 B.C. by inhabitants of the modern Middle East and Asia regions. A solar or electric food dehydrator can greatly speed the drying process and ensure more consistent results. Water is usually removed by evaporation (air drying, sun drying, smoking or wind drying) but, in the case of freeze-drying, food is first frozen and then the water is removed by sublimation. Bacteria, yeasts and molds need the water in the food to grow, and drying effectively prevents them from surviving in the food.

Many different foods are prepared by dehydration. Good examples are meat such as prosciutto (a.k.a. Parma ham), bresaola, and beef jerky. Dried and salted reindeer meat is a traditional Sami food. First the meat is soaked / pickled in saltwater for a couple of days to guarantee the conservation of the meat. Then the meat is dried in the sun in spring when the air temperature is below zero. The dried meat can be further processed to make soup.

Fruits change character completely when dried: the plum becomes a prune, the grape a raisin; figs and dates are also transformed in new, different products, that can be eaten as they are or else after rehydration. However, in dehydrating a grape into a raisin, antioxidants (Vitamin C, E, ORAC) and B vitamins are depleted, making the dehydrated fruit nutritionally inferior to its original source. Home drying of vegetables, fruit and even meat (to produce jerky) can be carried out in the home employing sun drying or using electrical dehydrators (household appliance). Preservatives such as potassium metabisulphite, or BHA, BHT for meat may be used. These preservatives are not required, however dried products without these preservatives may require refrigeration or freezing to ensure a long period of storage.

Freeze dried vegetables are often found in food for backpackers, hunters, the military, etc. Garlic and onion are often dried. Edible and psilocybin mushrooms, as well as other fungi, are also sometimes dried for preservation purposes, to affect the potency of chemical components, or so they can be used as seasonings.

For centuries, much of the European diet depended on dried cod, known as salt cod or bacalhau (with salt) or stockfish (without). It formed the main protein source for the slaves on the West Indian plantations, and was a major economic force within the triangular trade. Dried shark meat, known as Hákarl, is a delicacy in Iceland.

Grain drying

Hundreds of millions of tonnes of wheat, corn, soybean, rice and other grains as sorghum, sunflower seeds, rapeseed/canola, barley, oats, etc., are dried in grain dryers. In the main agricultural countries, drying comprises the reduction of moisture from about 17-30%w/w to values between 8 and 15%w/w, depending on the grain. The final moisture content for drying must be adequate for storage. The more oil the grain has, the lower its storage moisture content will be (though its initial moisture for drying will also be lower). Cereals are often dried to 14% w/w, while oilseeds, to 12.5% (soybeans), 8% (sunflower) and 9% (peanuts). Drying is carried out as a requisite for safe storage, in order to inhibit microbial growth. However, low temperatures in storage are also highly recommended to avoid degradative reactions and, especially, the growth of insects and mites.

A good maximum storage temperature is about 18°C. The largest dryers are normally used "Off-farm", in elevators, and are of the continuous type: Mixed-flow dryers are preferred in Europe, while Cross-flow dryers in the United States. In Argentina, both types are commonly found. Continuous flow dryers may produce up to 100 metric tonnes of dried grain per hour. The depth of grain the air must traverse in continuous dryers range from some 0.15 m in Mixed flow dryers to some 0.30 m in Cross-Flow. Batch dryers are mainly used "On-Farm", particularly in the United States and Europe. They normally consist of a bin, with heated air flowing horizontally from an internal cylinder through an inner perforated metal sheet, then through an annular grain bed, some 0.50 m thick (coaxial with the internal cylinder) in radial direction, and finally across the outer perforated metal sheet, before being discharged to the atmosphere. The usual drying times range from 1 h to 4 h depending on how much water must be removed, type of grain, air temperature and the grain depth. In the United States, continuous counterflow dryers may be found on-farm, adapting a bin to slowly drying grain fed at the top and removed at the bottom of the bin by a sweeping auger.

Grain drying is an active area of manufacturing and research. The performance of a dryer can be simulated with computer programmes based on mathematical models that represent the phenomena involved in drying: physics, physical chemistry, thermodynamics and heat and mass transfer. Most recently computer models have been used to predict product quality by achieving a compromise between drying rate, energy consumption, and grain quality. A typical quality parameter in wheat drying is breadmaking quality and germination percentage whose reductions in drying are somewhat related.

Grain Drying fundamentals

Drying starts at the bottom of the bin, which is the first place air contacts. The dry air is brought up by the fan through a layer of wet grain. Drying happens in a layer of 1 to 2 feet thick, which is called the drying zone. The drying zone moves from the bottom of the bin to the top, and when it reaches the highest layer, the grain is dry. The grain below drying zone is in equilibrium moisture content with drying air, which means it is safe for storage; while the grain above still needs drying. The air is then forced out the bin through exhaust vent.

Allowable Storage Time

Allowable storage time is an estimate of how long the grain needs to be dried before spoilage and maintain grain quality during storage. In grain storage process, fungi or molds are the primary concern. Many other factors, such as insects, rodents, and bacteria, also affect the condition of storage. The lower the grain temperature is, the longer the allowable storage time will be.

Proper moisture levels for safe storage

It is possible for long period safe storage if grain moisture content is less than 14%, and stored away from insects, rodents and birds. The following figure is the recommended moisture content for safe storage.

Storage duration	*Required MC for safe storage*	*Potential problems*
Weeks to a few months storage	14% or less	Molds, discoloration, respiration loss, insect damage, moisture adsorption
Storage for 8 to 12 months	13% or less	Insect damage
Storage of farmer's seeds	12% or less	Loss of germination
Storage for more than 1 year	9% or less	Loss of germination

Equilibrium Moisture Content

Moisture content in grain is related to the relative humidity and the temperature of the surrounding air. Equilibrium moisture content point is the point when grain no longer losing or gaining water when contacting with drying air. The final moisture content of the grain is up to the amount of moisture in the drying air, which is the relative humidity. The low relative humidity means air is dry and it has a large potential of picking up water. The lower the relative humidity is, the drier the air is. In general, one-half reduce in relative humidity is caused by 20° degree increase in air temperature.

Temperature

Heated air may be used in grain drying process. It can not only accelerate moisture migration inside the kernel, but also can evaporate the moisture on the surface. The major problem about heated air for drying process is the kernel temperature. The grain kernel may be damaged by high kernel temperatures. Usually, kernel temperature is lower than the air temperature. For different use of the corn, temperatures vary. For example, for seed corn, the maximum temperature is 110°F; for livestock feeding corn, the maximum temperature is 180°F.

Aeration

Aeration process refers to the process of moving air through grain. Airflow is a measurement of the amount of air in cubic feet per minute (CFM). In grain drying process, drying time is largely depended on aeration rates. Without sufficient airflow, grain maybe damaged before drying complete. Fans are used to move air through grain.

Application	*Grain aeration rate (cfm/bu)*
Quality maintenance	1/50 to 1
Natural-air bin drying	1 to 3
Heated-air bin drying	2 to 12
Batch or continuous-flow column dryers	50 to 150
Fluidization	~400

Drying cost

The drying cost is made up of two parts: the capital cost and the operating cost. Capital cost is largely depend on the drying rate requirement, and equipment cost. Operating cost refers to fuel, electricity and labour force cost. The amount of energy required to dry a bushel of grain is similar for all the drying methods. Some methods depend largely on natural air, while others may use LP heat or natural gas, which make energy cost vary. Basically, fuel and electrical power are the major portions of the operating cost. Drying cost is based on the B.T.U. consumption of temperature change from environment to desired one.

Classification of grain drying methods

In-storage drying methods

1. Low-temperature drying
2. Multiple-layer drying

Batch drying methods

1. Bin batch drying
2. Column batch drying

Continuous flow drying methods

1. Cross flow drying
2. Counter flow drying
3. Concurrent flow drying

In-storage Drying Methods

Low-temperature Drying: In-storage drying methods refers to those grain is dried and stored in the same container. Low-temperature drying, also known as near-ambient drying, is one of in-storage drying methods. There are four major factors which influence low temperature drying: the variability of weather, the harvest moisture content, the air flow in the storage bin and the amount of heated air. Most low-temperature dryers are built to dry grain as slowly as possible, while in the same time less spoilage on the grain. It is suggested that low temperature drying system is better operated when the average daily temperature is between 30°C and 50°C. Rather than control the drying air temperature, the low-temperature drying focuses on the relative humidity in order to achieve equilibrium moisture content (EMC) in all grain layers. Low-temperature drying process usually takes 5 days to several months depends on several important variables: weather, airflow, initial moisture content and amount of heat used. Among which, airflow is the key factor. Without appropriate airflow rate, spoilage will occur before drying is completed. By using heated air (LP heat, electric heat and solar heat), the relative humidity of the drying air is better controlled to achieve the desired moisture content. Usually, heated air dryer is used when the relative humidity larger than 70%. In electric heat dryers, an electrical resistant heater is usually placed before the fan to heat the airstream. In some case, a humidistat is employed to control the heater. In solar heat dryers, the drying air passes through the solar collector first to be heated (usually 10 to 12°F rise), then enters the bin through the fan and motor. The advantages of in-storage low temperature drying are quick filling, high quality product, less equipment requirement; while the disadvantages are long drying time, electrical demand if using electric heat, high management skills and uncertain harvest moisture content.

Multiple-layer Drying

Multiple layer drying method refers to the use of LP heat or natural gas in drying corn. Compared to low-temperature methods, multiple-layer drying requires higher temperatures, which results in a shorter allowable storage time. Multiple-layer drying without stirring is the basic multiple-layer drying method, in which airstream is entered through an LP heater by a fan. Usually, the temperature rise after the LP burner is remained low in order to avoid over drying in the bottom layers in the bin. As soon as corn is dried in the bin, the burner is turned off and the fan is used to bring the corn to ambient temperature. The advantages of multiple-layer drying without stirring are little handling of corn, and bin can be used as either dryer or storage; the disadvantages are slow filling and over drying in the bottom layers (Bern and Brumm, 2010). Multiple-layer drying with stirring can not only dry grain equilibrium from top to bottom, but also decrease the air resistance of the grain. Moreover, using stirring system can avoid over drying in bottom layer problem and give a uniform grain moisture content in the whole bin. When drying is complete, the burner is turned off while the fan and stirrer are used to mix the corn to achieve equal moisture content and temperature. The advantages of adding stirring are preventing over drying and accelerating drying and allowable fill rate; the disadvantages of stirring system are additional expenses and decreasing bin capacity.

Batch Drying Methods

Bin-Batch Drying: In batch drying methods, certain amount of grain is placed first, usually 2 to 4 inches, the batch is dried and cooled later, then drying is stopped and batch is removed. The batch dryers are usually operating under this sequence and repeating this sequence for several times. The bin-batch drying methods employ a full perforated floor as the dryer. Without stirring, large variety of equipment is available and the batch can be used as both dryer and cooler, but there may be large moisture gradient from top to bottom and losing time in loading and unloading process. When adding stirring system, unequilibrium moisture content problem is avoided, however, stirrer is an added expense. When using bin-batch roof dryer, time losing problem can be solved. There is a drying floor under the bin roof and the drying fan and burner is installed high on the bin wall. When the drying process is completed, grain is put in the regular bin floor, thus unloading time is reduced. However, there is no wet grain holding in bin-batch roof dryers and there is more expense on machines.

Column Batch Drying

The column formed in this kind of dryer is made up of two vertical perforated steel sheets, which is about 12 inches thick each. The capacity of column batch dryers is too small to store grains. The advantages of column batch, stationary bed dryer are easy to move and the dyer can be used as cooler; while the disadvantages are time losing when cooling, loading and unloading and unequal moisture distribution when drying is completed. When column batch recirculating dryer is used, the moisture content variation problem is avoided, but the additional handling process may result in grain spoilage.

Continuous Flow Drying Methods

Cross Flow Drying: Cross flow dryers is one of the most widely used continuous flow dryers. In the cross flow dryer, the airstream is perpendicular to the grain flow. Then the grain near the drying air is over dried, while on the other side, grain is under dried. Moisture gradient exists when drying is complete. In reality, the lower the airflow rate, the higher the grain moisture content variation between two sides of the column.

Concurrent Flow Drying

In the concurrent flow dryer, both the grain and air are moving in the same direction, which means the wettest grain is subjected to the hottest drying air. The kernels leave the drying region at the same temperature and the same moisture content. Energy efficiency is 40% better compared to cross flow dryer. However, the bed depth must be deeper than 12 inches depth than cross flow type. Thus, fan power requirements are high in this type of dryer.

Counter Flow Drying

In the counter flow dryer, the grain and the air are moving in opposite directions, which mean the driest grain is subjected to the hottest drying air. The kernels leave the drying region at the same temperature and the same moisture as in concurrent flow dryers. The suggesting air temperatures are less than 180°F because the driest kernels are more likely to be damaged by hot air.

Applications of Grain Drying

Sunflower Drying: For different types of sunflowers, the preservation moisture content is different. Oilseed sunflowers are better dried to 9 percent moisture content, while bird seed sunflowers are 10 percent moisture content. Compared to corn drying, sunflowers are more easily to be dried and kept in safe storage. What's more, high temperature may not have adverse effect on sunflowers kernel, which may be the reason of the fatty acid composition. There was no evidence of damage when air was heated up to 220°F in drying. However, fine hairs and fibres on the seed coat of sunflowers may cause fire hazard. It is suggested that remove the flaming particles first when hitting the sunflowers.

Bean Drying

The seed coat of bean is quite fragile and easy to damaged in cracking and splitting, which may cause loss to the producer. Some studies on beans suggested that in order to avoid cracking, it is better to keep drying air above 40 percent relative humidity.

Other Methods

Figure: *This electric food dehydrator has a hot air blower that blows air through trays with foods on them. Pictured are mango and papaya slices being dried.*

There are many different methods for drying, each with their own advantages for particular applications; these include:

- Bed dryers
- Drum drying
- Freeze Drying
- Shelf dryers
- Spray drying
- Sunlight
- Commercial food dehydrators
- Household oven

Food Dehydrator

Figure: *Tomato slices ready to be dried in a food dehydrator. In this model, multiple trays can be stacked on top of each other and warm air flows around the food.*

Figure: *Electric food dehydrator with mango and papaya slices being dried.*

Food drying is a way to preserve fruit, vegetables, and animal proteins after harvest, that has been practiced since antiquity, and a food dehydrator refers to a device that removes moisture from food to aid in its preservation. A food dehydrator uses a heat source and air flow to reduce the water content of foods. The water content of food is usually very high, typically 80% to 95% for various fruits and vegetables and 50% to 75% for various meats. Removing moisture from food restrains various bacteria from growing and spoiling food. Further, removing moisture from food dramatically reduces the weight of the food. Thus, food dehydrators are used to preserve and extend the shelf life of various foods.

Devices require heat using energy sources such as solar or electric power, and vary in form from large-scale dehydration projects to DIY projects or commercially sold appliances for domestic use. A commercial food dehydrator's basic parts usually consist of a heating element, a fan, air vents allowing for air circulation and food trays to lay food upon. A dehydrator's heating element, fans and vents simultaneously work to remove moisture from food. A dehydrator's heating element warms the food causing its moisture to be released from its interior. The appliance's fan then blows the warm, moist air out of the appliance via the air vents. This process continues for hours until the food is dried to a substantially lower water content, usually fifteen to twenty percent or less.

Most foods are dehydrated at temperatures of 130°F, or 54°C, although meats being made into jerky should be dehydrated at a higher temperature of 155°F, or 68°C, or preheated to those temperature levels, to guard against pathogens that may be in the meat. The key to successful food dehydration is the application of a constant temperature and adequate air flow. Too high a temperature can cause hardened foods: food that is hard and dry on the outside but moist, and therefore vulnerable to spoiling, on the inside.

Solar Food Dehydrators

Solar food drying involves the use of a solar dryer designed and built specifically for this purpose. Solar drying is distinctly different from open air "sun drying," which has been used for thousands of years. A good solar food dryer dries food much faster than air drying. It can achieve higher food drying temperature, control airflow and temperature, and keep food protected while drying.

Food drying is an excellent solar energy application since food drying primarily requires heat, and solar radiation is easily converted to heat. A clear or translucent glazing allows sunlight to enter an enclosed chamber where it is converted to heat when it strikes a dark interior surface. Airflow is typically achieved with natural convection (warm air rises). Adjustable venting allows regulation of airflow and temperature.

Solar food drying is effective and practical in most of the populated places of the world. A general rule is that, if you can grow a successful vegetable garden, then there is enough solar energy to dry the food you produce (some overcast, northern maritime climates are the exception).

Some solar food dryer designs employ a separate solar collector to generate the heated air, which is then directed into a food chamber or cabinet. Other designs combine the collector and food cabinet and allow direct heating of food. Backup electric heating can be incorporated into some solar food dehydrators to provide an alternative heat source if the weather changes.

Solar food dehydrators are often cited as viable tools in the search for agricultural sustainability and food security.

Heat and Mass Transfer

Many food and agricultural processes involve heat and mass transfer processes. Mathematical modelling and computer simulation of grain drying are now widely used in agricultural engineering research. Several models have been proposed to describe the heat and mass transfer processes in the basic types of convective grain drier. Most of these models, however, have been derived under assumptions which are not explicitly stated and which restrict their applications from the outset. Furthermore, the differences which exist between various models are not always clarified in the literature. It is important with an ever-increasing demand for the accurate modelling of drying systems, for the researcher to understand the basic assumptions inherent in a particular model and hence to be aware of its limitations. In addition, the problems of obtaining satisfactory solutions for particular models have generally been given only a cursory treatment.

Drying, freezing, de-frosting and packing as well as storage are in the group of technological processes of vegetable processing during which the raw material exchanges heat and mass with the environment. Since the energy outlays needed to carry out these processes are usually high (e.g. convective drying, cooling, storing) it is important of optimise them in order to reduce the costs of obtaining a high-quality product. The optimal conditions during the process may be determined effectively and cheaply using the standard methods of the process optimisation. Those methods make us use mathematical models describing the processes under investigation. Among the many available types of mathematical models of food processing, models formed from differential equations may be distinguished. According to the character of changes of analysed values, we may distinguish models with lumped parameters as well as the models with distributed parameters. The models with distributed parameters are made up of partial differential equations and make it possible to determine changes in values depending both on time and location (e.g. change of concentration of phenolic compounds in apple slices according to the time of treatment and location within the slice). Such models are usually created using the equations of mass, momentum and energy balances applied to a single particle or the entire bed of particles of the raw material being processed. The models used in the process of food engineering are usually highly non-linear and the simulation results are usually obtained on the basis of the finite elements method, and less frequently on the basis of the finite differences method.

Among many problems connected with the development and study of models with distributed parameters as applied to food processing the problem of identification of parameters of the model is worth stressing. It appears that the models of food processing (including food storage) are very often susceptible to changes of parameters. This equally applies to balance equations as well as initial and boundary conditions

Agricultural Machinery

With the coming of the Industrial Revolution and the development of more complicated machines, farming methods took a great leap forward. Instead of harvesting grain by hand with a sharp blade, wheeled machines cut a continuous swath. Instead of threshing the grain by beating it with sticks, threshing machines separated the seeds from the heads and stalks. The first tractors appeared in the late 19th century.

Steam Power

Power for agricultural machinery was originally supplied by horses or other domesticated animals. With the invention of steam power came the portable engine, and later the traction engine, a multipurpose, mobile energy source that was the ground-crawling cousin to the steam locomotive. Agricultural steam engines took over the heavy pulling work of horses, and were also equipped with a pulley that could power stationary machines via the use of a long belt. The steam-powered machines were low-powered by today's standards but, because of their size and their low gear ratios, they could provide a large drawbar pull. Their slow speed led farmers to comment that tractors had two speeds: "slow, and damn slow."

Internal Combustion Engines

The internal combustion engine; first the petrol engine, and later diesel engines; became the main source of power for the next generation of tractors. These engines also contributed to the development of the self-propelled, combined harvester and thresher, or combine harvester (also shortened to 'combine'). Instead of cutting the grain stalks and transporting them to a stationary threshing machine, these combines cut, threshed, and separated the grain while moving continuously through the field.

Types

Figure: *A John Deere cotton harvester at work in a cotton field.*

Combines might have taken the harvesting job away from tractors, but tractors still do the majority of work on a modern farm. They are used to pull implements—machines that till the ground, plant seed, and perform other tasks.

Tillage implements prepare the soil for planting by loosening the soil and killing weeds or competing plants. The best-known is the plow, the ancient implement that was upgraded in 1838 by John Deere. Plows are now used less frequently in the U.S. than formerly, with offset disks used instead to turn over the soil, and chisels used to gain the depth needed to retain moisture. The most common type of seeder is called a planter, and spaces seeds out equally in long rows, which are usually two to three feet apart. Some crops are planted by drills, which put out much more seed in rows less than a foot apart, blanketing the field with crops. Transplanters automate the task of transplanting seedlings to the field. With the widespread use of plastic mulch, plastic mulch layers, transplanters, and seeders lay down long rows of plastic, and plant through them automatically.

After planting, other implements can be used to cultivate weeds from between rows, or to spread fertilizer and pesticides. Hay balers can be used to tightly package grass or alfalfa into a storable form for the winter months.

Modern irrigation relies on machinery. Engines, pumps and other specialized gear provide water quickly and in high volumes to large areas of land. Similar types of equipment can be used to deliver fertilizers and pesticides.

Besides the tractor, other vehicles have been adapted for use in farming, including trucks, airplanes, and helicopters, such as for transporting crops and making equipment mobile, to aerial spraying and livestock herd management.

New technology and the future

The basic technology of agricultural machines has changed little in the last century. Though modern harvesters and planters may do a better job or be slightly tweaked from their predecessors, the US$250,000 combine of today still cuts, threshes, and separates grain in essentially the same way it has always been done. However, technology is changing the way that humans operate the machines, as computer monitoring systems, GPS locators, and self-steer programmes allow the most advanced tractors and implements to be more precise and less wasteful in the use of fuel, seed, or fertilizer. In the foreseeable future, there may be mass production of driverless tractors, which use GPS maps and electronic sensors. Even more esoteric are the new areas of nanotechnology and genetic engineering, where submicroscopic devices and biological processes, respectively, are being used as machines to perform agricultural tasks in unusual new ways.

Agriculture may be one of the oldest professions, but the development and use of machinery has made the job title of *farmer* a rarity. Instead of every person having to work to provide food for themselves, less than 2% of the U.S. population today works in agriculture, yet that 2% provides considerably more food than the other 98% can eat. It is estimated that at the turn of the 20th century, one farmer in the U.S. could feed 25 people, where today, that ratio is 1:130 (in a modern grain farm, a single farmer can produce cereal to feed over a thousand people). With continuing advances in agricultural machinery, the role of the farmer continues on.

The next advance in farming will be the electrification of agricultural machines to improve energy efficiency reduce the environmental energy budget. While an all electric tractor with fully electric agricultural machines may still be several years out, companies are now looking to provide mobile power sources for agricultural applications. The Power Pack 45 from Raussendorf could help fuel the trend to electrify agricultural machines.

2

Agricultural Structures

Agricultural buildings are structures designed for farming and agricultural practices, including but not limited to: growing and harvesting of crops and raising livestock and small animals.

Specific examples of agricultural buildings include:

- barns
- greenhouses
- storage buildings for farm equipment, animal supplies or feed
- storage buildings for equipment used to implement farming and/or agricultural practices
- storage buildings for crops grown and raised on site (cold storage)
- horticultural nursery.

Grain Storage Structures

In most countries grains are among the most important staple foods. However they are produced on a seasonal basis, and in many places there is only one harvest a year, which itself may be subject to failure. This means that in order to feed the world's population, most of the global production of maize, wheat, rice, sorghum and millet must be held in storage for periods varying from one month up to more than a year. Grain storage therefore occupies a vital place in the economies of developed and developing countries alike.

The market for food grains is characterized by fairly stable demand throughout the year, and widely fluctuating supply. Generally speaking people's consumption of basic foods such as grains does not vary greatly from one season to another or from year to year. The demand for grain is 'inelastic', which means that large changes in the market price lead to relatively small changes in the amount of grains which people purchase.

Market supply, on the other hand, depends on the harvest of grains which is concentrated within a few months of the year in any one area, and can fluctuate widely from one year to the next depending on climatic conditions. New varieties that have shorter growing periods, and variation in climatic conditions and farming systems in different regions of a country, can help to even out the fluctuations in market supply. But even in a country such as Indonesia, which has diverse climatic and farming conditions and where 90 per cent of rice land is under short duration high yielding varieties, about 60 per cent of production is harvested within a three month period (Ellis et al. 1992).

The main function of storage in the economy is to even out fluctuations in market supply, both from one season to the next and from one year to the next, by taking produce off the market in surplus seasons, and releasing it back onto the market in lean seasons. This in turn smooths out out fluctuations in market prices. The desire to stabilise prices of basic foods is one of the major reasons why governments try to influence the amount of storage occurring, and often undertake storage themselves.

Both producers and consumers benefit from stable prices, which reduce the uncertainties associated with planning farm investment and household expenditure. However storage involves costs, and the only way in which these costs can be recuperated is through a price spread. If storage is to be profitable, people who store grain must receive a price on sale which at least covers the costs of storing the grain since harvest. These include:

- The cost of the store itself (often a rental cost);
- Labour and supervision;
- Pest control;
- Storage and spillage losses; and
- Cost of capital invested in the grain.

In practice, the costs of storage depend on the commodity stored, on the type of storage system, and on unpredictable and variable factors such as pest incidence and climatic conditions. Storage costs also depend on the circumstances of the person, the business or the institution who is storing. The most variable component of storage costs is the cost of capital. For a small farmer or trader, capital may be scarce and costly, and their only access to loans may be from money lenders charging rates of 10% or more per month. On the other hand a Government Marketing Board may have preferential access to loans at low interest, at rates of as low as 10% per annum. There is, therefore, no single cost of storage.

Until now, most large-scale storage of cereals in developing countries has been carried out by Government marketing boards, which have developed their activities with the technical and financial assistance of donors and international financial institutions. This has resulted in the building of large numbers of grain stores and mills, with a gradual increase in scale and the degree of technical sophistication. In some countries this has helped maintain food security and feed urban populations growing at 5% or more per annum. However there has also been much wastage, with stores often being inadequately located, inappropriately designed, and poorly managed and maintained. Stores have been built to support Government monopolies, on the assumption that no storage would be carried out by the private sector, but in the event of liberalisation, much of the capacity has been found unneeded and poorly located. Thus the National Milling Corporation in Tanzania has over 400,000 tonnes of storage capacity for which it must find new uses, to which must be added the large storage capacity of the State-sponsored cooperative unions.

Most grain stores have been designed for bag handling, reflecting the low cost of manual labour and the lack of spare parts and maintenance support needed for mechanical handling equipment. However bulk handling has been introduced throughout the developing world, with results which leave much to be desired. Too often, Marketing Boards have been unable to make good use of these facilities and they have become rusting monuments to inappropriate development assistance.

In international development assistance programmes, there is often much support for 'modern' capital-intensive systems. This was observed in a study commissioned by the Pakistan Agricultural Research Council (Courter, 1991). Between 1983 and 1987 no less than five feasibility studies advanced the case for investments in bulk handling. All the studies had used questionable assumptions, and four out of the five studies had assumed reductions in storage losses of 5% or more simply by switching from bag to bulk handling. Given that loss surveys have revealed storage losses of between 1.5% and 3.9% such reductions would have been impossible.

The Case for Bulk Handling

There are, however, instances when bulk handling is economically justified in developing countries. This is usually at bottle-necks in the marketing chain, and where grain has to be handled in large volumes and at great speed, for example at port facilities, railway terminals or at mills. In such cases, investment in bulk handling facilities will often result in major cost savings, due to reduction in demurrage charges and down-time. The investment in one capital asset (bulk-handling equipment) produces major savings in the use of other capital assets (ships, trains, mills etc.). In most developing countries, labour costs are unlikely to be significant in the calculations, but bulk handling will reduce the risk that labour problems, strikes etc. will slow operations or bring them to a halt.

Bulk handling tends to be least viable for the long-term storage of grain, where stocks are only turned over once a year or less. In such cases the savings in labour and bags are unlikely to cover the high capital cost in silos, handling equipment etc. With well-run bulk storage complexes it may be possible to achieve a marginal reduction in storage losses but, due to poor operation and maintenance, losses are sometimes increased.

Bulk handling may also result from changes in farming methods. The introduction of combine harvesters in certain countries provides an incentive to start handling the grain in bulk. From the tank of the harvester the grain can be transferred mechanically to a bulk truck, from there to a bulk store or silo, and from there to a mill. This completely eliminates the use of the bags in the system, and overcomes problems of labour shortage and congestion which sometimes occur with a bag handling system. However in irrigated or high rainfall areas, the main benefit to the farmer may be by making it easier for him/her to plant a second crop on the same land.

Silo

A silo (from the Greek óéñüò – *siros*, "pit for holding grain") is a structure for storing bulk materials. Silos are used in agriculture to store grain or fermented feed known as silage. Silos are more commonly used for bulk storage of grain, coal, cement, carbon black, woodchips, food products and sawdust. Three types of silos are in widespread use today: tower silos, bunker silos, and bag silos.

Silo Types

There are different types of cement silos such as the low-level mobile silo and the static upright cement silo, which are used to hold and discharge cement and other powder materials such as PFA (Pulverised Fuel Ash). The low-level silos are fully mobile with capacities from 10 to 75 tons. They are simple to transport and are easy to set up on site. These mobile silos generally come equipped with an electronic weighing system with digital display and printer. This allows any quantity of cement or powder discharged from the silo to be controlled and also provides an accurate indication of what remains inside the silo. The static upright silos have capacities from 20 to 80 tons. These are considered a low-maintenance option for the storage of cement or other powders. Cement silos can be used in conjunction with bin-fed batching plants.

Cement storage silos:

Figure: *Coal silo under construction using aluminum concrete formwork*

Tower Silo

Storage silos are cylindrical structures, typically 10 to 90 ft (4 to 30 m) in diameter and 30 to 275 ft (10 to 84 m) in height with the slipform and Jumpform concrete silos being the larger diameter and taller silos. They can be made of many materials. Wood staves, concrete staves, cast concrete, and steel panels have all been used, and have varying cost, durability, and airtightness tradeoffs. Silos storing grain, cement and woodchips are typically unloaded with air slides or augers. Silos can be unloaded into rail cars, trucks or conveyors.

Tower silos containing silage are usually unloaded from the top of the pile, originally by hand using a silage fork, which has many more tines than the common pitchfork, 12 vs 4, in modern times using mechanical unloaders. Bottom silo unloaders are utilized at times but have problems with difficulty of repair.

An advantage of tower silos is that the silage tends to pack well due to its own weight, except in the top few feet. However, this may be a disadvantage for items like chopped wood. The tower silo was invented by Franklin Hiram King.

In Canada, Australia and the United States, many country towns or the larger farmers in grain-growing areas have groups of wooden or concrete tower silos, known as grain elevators, to collect grain from the surrounding towns and store and protect the grain for transport by train, truck or barge to a processor or to an export port. In bumper crop times, the excess grain is stored in piles without silos or bins, causing considerable losses.

Concrete Stave Silos

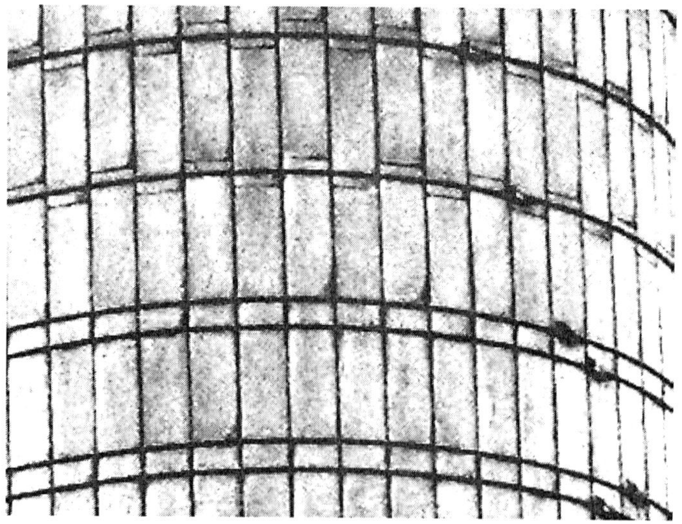

Figure: *High contrast image showing the intermeshed concrete staves, and how the lower hoops are aligned over the stave edges.*

Figure: *Small stave silos can be expanded upward. More hoops are added to strengthen the lower staves.*

Concrete stave silos are constructed from small precast concrete blocks with ridged grooves along each edge that lock them together into a high strength shell. Much of concrete's strength comes from its high incompressibility, so the silo is held together by steel hoops encircling the tower and compressing the staves into a tight ring. The vertical stacks are held together by intermeshing of the ends of the staves by a short distance around the perimeter of each layer, and hoops which are tightened directly across the stave edges.

The static pressure of the material inside the silo pressing outward on the staves increases towards the bottom of the silo, so the hoops can be spaced wide apart near the top but become progressively more closely spaced towards the bottom to prevent seams from opening and the contents leaking out.

Concrete stave silos are built from common components designed with high strength and long life. They have the flexibility to have their height increased according to the needs of the farm and purchasing power of the farmer, or to be completely disassembled and reinstalled somewhere else if no longer needed.

Low-oxygen Tower Silos

Low-oxygen silos are designed to keep the contents in a low-oxygen atmosphere at all times to keep the fermented contents in a high quality state and to prevent mold and decay, as may occur in the top layers of a stave silo or bunker. Low-oxygen silos are only opened directly to the atmosphere during the initial forage loading, and even the unloader chute is sealed against air infiltration.

Figure: *Low-oxygen Harvestore tower silos*

It would be expensive to design a such a large structure that is immune to atmospheric pressure changes over time. Instead, the silo structure is open to the atmosphere but outside air is separated from internal air by large impermeable bags sealed to the silo breather openings. In the warmth of the day when the silo is heated by the sun, the gas trapped inside the silo expands and the bags "breathe out" and collapse. At night the silo cools, the air inside contracts and the bags "breathe in" and expand again.

While the iconic blue Harvestore low-oxygen silos were once very common, the speed of its unloader mechanism was not able to match the output rates of modern bunker silos, and this type of silo went into decline. Unloader repair expenses also severely hurt the Harvestore reputation, because the unloader feed mechanism is located in the bottom of the silo under tons of silage. In the event of cutter chain breakage, it can cost up to US$10,000 to perform repairs. The silo may need to be partially or completely emptied by alternate means, to unbury the broken unloader retrieve broken components lost in the silage at the bottom of the structure.

In 2005 the Harvestore company recognized these issues and worked to develop new unloaders with double the flow rate of previous models to stay competitive with bunkers, and with far greater unloader chain strength. They are now also using load sensing soft-start variable frequency drive motor controllers to reduce the likelihood of mechanism breakage, and to control the feeder sweep arm movement.

Figure: *This bunker silo contains sugar. Chillan, Chile*

While the sight of multiple tall silos may look impressive, many have been abandoned for silage use, or converted to hold grain. Ground-level bunker silos carry none of the hazards outlined above, and need only a standard tractor/loader to feed out with, and no specialized machinery.

Bunker Silos

Bunker silos are trenches, usually with concrete walls, that are filled and packed with tractors and loaders. The filled trench is covered with a plastic tarp to make it airtight. These silos are usually unloaded with a tractor and loader. They are inexpensive and especially well suited to very large operations.

Bag Silos

Bag silos are heavy plastic tubes, usually around 8 to 12 ft in diameter, and of variable length as required for the amount of material to be stored. They are packed using a machine made for the purpose, and sealed on both ends. They are unloaded using a tractor and loader or skid-steer loader. The bag is discarded in sections as it is torn off. Bag silos require little capital investment. They can be used as a temporary measure when growth or harvest conditions require more space, though some farms use them every year.

Figure: *8' diameter by 150 foot silo bag shown just after filling and sealing.*

Bins

Figure: *This silo bin contains 27 variations of stone, sand and gravel, Copenhagen, Denmark*

A bin is typically much shorter than a silo, and is typically used for holding dry matter such as concrete or grain. Bins may be round or square, but round bins tend to empty more easily due to a lack of corners for the stored material to become wedged and encrusted.

The stored material may be powdered, as seed kernels, or as cob corn. Due to the dry nature of the stored material, it tends to be lighter than silage and can be more easily handled by under-floor grain unloaders. To facilitate drying after harvesting, some grain bins contain a hollow perforated or screened central shaft to permit easier air infiltration into the stored grain.

Sand and Salt Silos

Sand and salt for winter road maintenance are stored in conical dome-shaped silos. These are more common in North America, namely in Canada and the United States.

History

Archaeological ruins and ancient texts show that silos were used in ancient Greece as far back as the late 8th century BC, as well as the 5th Millennium B.C site of Tell Tsaf, Israel. The term *silo* is derived from the Greek óéñüò (*siros*), "pit for holding grain".

The first modern silo, a wooden and upright one filled with grain, was invented and built in 1873 by Fred Hatch of McHenry County, Illinois, USA.

Forage Silo Usage

Forage Harvesting

Forage silo filling is performed using a forage harvester which may either be self-propelled with an engine and driver's cab, or towed behind a tractor that supplies power through a PTO.

The harvester contains a drum-shaped series of cutting knives which shear the fibrous plant material into small pieces no more than an inch long, to facilitate mechanized blowing and transport via augers. The finely chopped plant material is then blown by the harvester into a forage wagon which contains an automatic unloading system.

Tower Filling

Figure: *Short video of the steps involved for filling a farm tower silo, with English captions. 4 min 15 sec 320x240 30fps 200 kbit Low quality*

Tower forage filling is typically performed with a silo blower which is a very large fan with paddle-shaped blades. Material is fed into a vibrating hopper and is pushed into the blower using a spinning spiral auger.

There is commonly a water connection on the blower to add moisture to the plant matter being blown into the silo. The blower may be driven by an electric motor but it is more common to use a spare tractor instead.

A large slow-moving conveyor chain underneath the silage in the forage wagon moves the pile towards the front, where rows of rotating teeth break up the pile and drop it onto a high-speed transverse conveyor that pours the silage out the side of the wagon into the blower hopper.

Bag Filling

Silo bags are filled using a travelling sled driven from the PTO of a tractor left in neutral and which is gradually pushed forward as the bag is filled. The steering of the tractor controls the direction of bag placement as it fills, but bags are normally laid in a straight line.

The bag is loaded using the same forage harvesting methods as the tower, but the forage wagon must be moved progressively forward with the bag loader. The loader uses an array of rotating cam-shaped spiraled teeth associated with a large comb-shaped tines to push forage into the bag. The forage is pushed in through a large opening, and as the teeth rotate back out, they pass between the comb tines. The cam-shaped auger teeth essentially wipe the forage off using the steel tines, keeping the forage in the bag.

Before filling begins, the entire bag is placed onto the loader as a bunched-up tube folded back on itself in many layers to form a thick pile of plastic. Because the plastic is minimally elastic, the loader mechanism filling chute is slightly smaller than the final size of the bag, to accommodate this stack of plastic around the mouth of the loader. The plastic slowly unfurls itself around the edges of the loader as the tube is filled.

The contents of the silo bag are under pressure as it is filled, with the pressure controlled by a large brake shoe pressure regulator, holding back two large winch drums on either side of the loader. Cables from the drum extend to the rear of the bag where a large mesh basket holds the rear end of the bag shut.

To prevent molding and to assure an airtight seal during fermentation, the ends of the silo bag tube are gathered, folded, and tied shut to prevent oxygen from entering the bag. Removal of the bag loader can be hazardous to bystanders since the pressure must be released and the rear end allowed to collapse onto the ground.

Tower Unloading

A *silo unloader* specifically refers to a special cylindrical rotating forage pickup device used inside a single tower silo.

The main operating component of the silo unloader is suspended in the silo from a steel cable on a pulley that is mounted in the top-centre of the roof of the silo. The vertical positioning of the unloader is controlled by an electric winch on the exterior of the silo.

For the summer filling of a tower silo, the unloader is winched as high as possible to the top of the silo and put into a parking position. The silo is filled with a silo blower, which is literally a very large fan that blows a large volume of pressurized air up a 10-inch tube on the side of the silo. A small amount of water is introduced into the air stream during filling to help lubricate the filling tube. A small adjustable nozzle at the top, controlled by a handle at the base of the silo directs the silage to fall into the silo on the near, middle, or far side, to facilitate evenly layered loading. Once completely filled, the top of the exposed silage pile is covered with a large heavy sheet of silo plastic which seals out oxygen and permits the entire pile to begin to ferment in the autumn.

In the winter when animals must be kept indoors, the silo plastic is removed, the unloader is lowered down onto the top of the silage pile, and a hinged door is opened on the side of the silo to permit the silage to be blown out. There is an array of these access doors arranged vertically up the side of the silo, with an unloading tube next to the doors that has a series of removable covers down the side of the tube. The unloader tube and access doors are normally covered with a large U-shaped shield mounted on the silo, to protect the farmer from wind, snow, and rain while working on the silo.

The silo unloader mechanism consists of a pair of counter-rotating toothed augers which rip up the surface of the silage and pull it towards the centre of the unloader. The toothed augers rotate in a circle around the centre hub, evenly chewing the silage off the surface of the pile. In the centre, a large blower assembly picks up the silage and blows it out the silo door, where the silage falls by gravity down the unloader tube to the bottom of the silo, typically into an automated conveyor system.

The unloader is typically lowered only a half-inch or so at a time by the operator, and the unloader picks up only a small amount of material until the winch cable has become taut and the unloader is not picking up any more material. The operator then lowers the unloader another half-inch or so and the process repeats. If lowered too far, the unloader can pull up much more material than it can handle, which can overflow and plug up the blower, outlet spout, and the unloader tube, resulting in a time-wasting process of having to climb up the silo to clear the blockages.

Once silage has entered the conveyor system, it can be handled by either manual or automatic distribution systems. The simplest manual distribution system uses a sliding metal platform under the pickup channel. When slid open, the forage drops through the open hole and down a chute into a wagon, wheelbarrow, or open pile. When closed, the forage continues past the opening and onward to other parts of the conveyor. Computer automation and a conveyor running the length of a feeding stall can permit the silage to be automatically dropped from above by each animal, with the amount dispensed customized for each location.

Safety

Figure: *Defunct silo in Merrinee, Victoria, Australia.*

Silos are hazardous, and people are killed or injured every year in the process of filling and maintaining them. The machinery used is dangerous and with tower silos workers can fall from the silo's ladder or work platform. Several fires have occurred over the years.

Dangers of Loading Process

Filling a silo requires parking two tractors very close to each other, both running at full power and with live PTO shafts, one powering the silo blower and the other powering a forage wagon unloading fresh-cut forage into the blower. The farmer must continually move around in this highly hazardous environment of spinning shafts and high-speed conveyors to check material flows and adjust speeds, and to start and stop all the equipment between loads.

Preparation for filling a silo requires winching the unloader to the top, and any remaining forage at the base that the unloader could not pick up must be removed from the floor of the silo. This job requires that the farmer work directly underneath a machine weighing several tons suspended fifty feet or more overhead from a small steel cable. If the unloader were to fall, the farmer would likely be killed instantly.

Dangers of Unloading Process

Unloading also poses its own special hazards, due to the requirement that the farmer regularly climb the silo to close an upper door and open a lower door, moving the unloader chute from door to door in the process. The fermentation of the silage produces methane gas which over time will outgas and displace the oxygen in the top of the silo. A farmer directly entering a silo without any other precautions can be asphyxiated by the methane, knocked unconscious, and silently suffocate to death before anyone else knows what has happened. It is either necessary to leave the silo blower attached to the silo at all times to use it when necessary to ventilate the silo with fresh air, or to have a dedicated electric fan system to blow fresh air into the silo, before anyone attempts to enter it.

In the event that the unloader mechanism becomes plugged, the farmer must climb the silo and directly stand on the unloader, reaching into the blower spout to dig out the soft silage. After clearing a plug, the forage needs to be forked out into an even layer around the unloader so that the unloader does not immediately dig into the pile and plug itself again. All during this process the farmer is standing on or near a machine that could easily kill them in seconds if it were to accidentally start up. can happen if someone in the barn were to unknowingly switch on the unloading mechanism while someone is in the silo working on the unloader.

Often, when unloading grain from an auger or other opening at the bottom of the silo, another worker will be atop the grain "walking it down", to ensure an even flow of grain out of the silo. Sometimes unstable pockets in the grain will collapse beneath the worker doing the walking; this is called grain entrapment as the worker can be completely sunk into the grain within seconds. Entrapment can also occur in moving grain, or when workers clear large clumps of grain that have become stuck on the side of the silo. This often results in death by suffocation.

Dry-material / Bin Hazards

There have also been many cases of silos and the associated ducts and buildings exploding. If the air inside becomes laden with finely granulated particles, such as grain dust, a spark can trigger an explosion powerful enough to blow a concrete silo and adjacent buildings apart, usually setting the adjacent grain and building on fire. Sparks are often caused by (metal) rubbing against metal ducts; or due to static electricity produced by dust moving along the ducts when extra dry.

The two main problems which will necessitate cleaning in dry-matter silos and bins are *bridging* and *rat-holing*. Bridging occurs when the material interlaces over the unloading mechanism at the base of the silo and blocks the flow of stored material by gravity into the unloading system. Rat-holing occurs when the material starts to adhere to the side of the silo. This will reduce the operating capacity of a silo as well as leading to cross-contamination of newer material with older material. There are a number of ways to clean a silo and many of these carry their own risks. However since the early 1990s acoustic cleaners have become available. These are non-invasive, have minimum risk, and can offer a very cost-effective way to keep a small particle silo clean.

Notable Silos

- Henninger Turm, Frankfurt, Germany, has an observation deck and 2 revolving restaurants, height: 120 metres
- Schapfen-Mill-Tower, Ulm, Germany, height: 115 metres
- Silo Tower Basel, Basel, Switzerland, has an observation deck, height: 52 metres
- Quaker Square, Akron, Ohio, United States, is a former set of tower silos that is now a hotel, restaurants and shops
- *Dagon*, Haifa, Israel - transformed into a museum of agriculture, a prominent local feature.

Granary

Figure: *Ancient Greek geometric art box in the shape of granaries, 850 BC. On display in the Ancient Agora Museum in Athens, housed in the Stoa of Attalos.*

Figure: *Granary in Kashan, Iran*

A granary is a storehouse or room in a barn for threshed grain or animal feed. Ancient or primitive granaries are most often made out of pottery. Granaries are often built above the ground to keep the stored food away from mice and other animals.

Early Origins

From ancient times grain has been stored in bulk. The oldest granaries yet found date back to 9500 BC and are located in the Pre-Pottery Neolithic A settlements in the Jordan Valley. The first were located in places between other buildings. However beginning around 8500 BC, they were moved inside houses, and by 7500 BC storage occurred in special rooms. The first granaries measured 3 x 3 m on the outside and had suspended floors that protected the grain from rodents and insects and provided air circulation.

These granaries are followed by those in Mehrgarh in the Indus Valley from 6000 BC. The ancient Egyptians made a practice of preserving grain in years of plenty against years of scarcity. The climate of Egypt being very dry, grain could be stored in pits for a long time without discernible loss of quality. The silo pit, as it has been termed, has been a favourite way of storing grain from time immemorial in all oriental lands. In Turkey and Persia, usurers used to buy up wheat or barley when comparatively cheap, and store it in hidden pits against seasons of dearth. In Malta a relatively large stock of wheat was preserved in some hundreds of pits (silos) cut in the rock. A single silo stored from 60 to 80 tons of wheat, which, with proper precautions, kept in good condition for four years or more.

East Asia

Simple storage granaries raised up on four or more posts appeared in the Yangshao culture in China and after the onset of intensive agriculture in the Korean peninsula during the Mumun pottery period (c. 1000 B.C.) as well as in the Japanese archipelago during the Final Jômon/Early Yayoi periods (c. 800 B.C.). In the archaeological vernacular of Northeast Asia, these features are lumped with those that may have also functioned as residences and together are called 'raised floor buildings'.

In vernacular architecture of Indonesian archipelago granaries are made of wood and bamboo materials and most of them are built raised up on four or more posts to avoid rodents and insects. Examples of Indonesian granary is Sundanese *leuit* and Minang *rangkiang*.

Modern

Towards the close of the 19th century, warehouses specially intended for holding grain began to multiply in Great Britain, but North America is the home of great granaries, known there as grain elevators. There are climatic difficulties in the way of storing grain in Great Britain on a large scale, but these difficulties have been largely overcome. To preserve grain in good condition it must be kept away as much as possible from moisture and heat. New grain when brought into a warehouse has a tendency to release moisture. Bacteria are more active in this condition and can heat the grain. If the heating is allowed to continue the quality of the grain suffers. An effectual remedy is to turn out the grain in layers, not too thick, on a floor, and to keep turning it over so as to aerate it thoroughly. Grain can thus be conditioned for storage in silos and they also used warehouse.

In Great Britain small granaries were built on mushroom shaped stumps called staddle stones. They were built of timber frame construction and often had slate roofs. Larger ones were similar to linhays, but with the upper floor enclosed. Access to the first floor was usually via stone staircase on the outside wall.

Storage For Semi-Perishable Products

Proper food storage helps maintain food quality by retaining flavour, colour, texture and nutrients, while reducing the chance of contracting a food-borne illness.

Foods can be classified into three groups. See a chart of storage life by following each link:

- Perishable foods includes meat, poultry, fish, milk, eggs and many raw fruits and vegetables. All cooked foods are considered perishable foods. To store these foods for any length of time, perishable foods need to be held at refrigerator or freezer temperatures. If refrigerated, perishable foods should be used within several days.
- Semi-perishable foods, if properly stored and handled, may remain unspoiled for six months to about one year. Flour, grain products, dried fruits and dry mixes are considered semi-perishable.
- Staple, or non-perishable, foods such as sugar, dried beans, spices and canned goods do not spoil unless they are handled carelessly. These foods will lose quality, however, if stored over a long time, even if stored under ideal conditions.

There is no exact method to determine how long a food will maintain quality and be safe to eat, because many conditions affect quality. The storage life of foods is affected by the:

- freshness of the food when it reached the grocery store

- length of time and the temperature at which it was held before purchase
- temperature of your food storage areas
- humidity level in your food storage areas
- type of storage container or packaging the food is stored in
- characteristics of the food item

Animal Shelters

Farm animals are domestic animals, they cannot "fend for themselves" on the range, or find their own housing. Appropriate shelter must be provided for all farm animals and includes: 1) Well-ventilated; 2) Weatherproof; and 3) Predator-proof housing.

Ventilation: Good ventilation is extremely important for all farm animal species. Though most important in hot weather, farm animals also need good ventilation in colder seasons. During the winter months, don't overheat barns or close your barn entirely. Humidity from urine, manure, and body moisture may arise and result in pneumonia. The most caring farm animal caregiver makes this easy mistake in an attempt to keep the animals in an environment that is "comfortable." Also, just because it's cold or snowing outside, don't necessarily assume the animals want to be kept indoors. Farm animals need (and prefer) fresh air, and generally go outdoors in even the harshest weather. For this reason, we feel it is important to arrange your barn and fencing so that the animals can choose to go inside or outside during the day.

Weatherproofing: All farm animal shelters must be weatherproof. In addition to being uncomfortable for the animals, wet, soiled bedding is a health hazard. During cold weather months, you will also need to provide the animals with extra straw for insulation and warmth.

Predator-proofing: The specific type of predator-proofing you will need depends on the type of farm animal under your care, but all need protection. Fencing is a must to keep predators out and farm animals in. Sadly, dogs are one of the most common farm animal predators. If you have dog companions in your family, carefully introduce them to farm animals and do not leave them alone together until you are absolutely sure it is safe. Cats and farm animals get along fine (except for baby birds or rabbits). Never put a farm animal on a rope or chain and leave him or her unprotected. The practice of "tethering" is stressful for the animals, and dangerous.

Material Handling Equipments

Material handling equipment is equipment that relate to the movement, storage, control and protection of materials, goods and products throughout the process of manufacturing, distribution, consumption and disposal. Material handling equipment is the mechanical equipment involved in the complete system. Material handling equipment is generally separated into four main categories: storage and handling equipment, engineered systems, industrial trucks, and bulk material handling.

Material Handling and Efficiency

Material handling equipment is used to increase output, control costs, and maximize productivity. There are several ways to determine if the material handling equipment is achieving peak efficiency. These include capturing all relevant data related to the warehouse's operation, measuring how many times an item is "touched" from the time it is ordered until it leaves the building, making sure you are using the proper picking technology, and keeping system downtime to a minimum. A special analytical data-set known as Stock-keeping units (SKUs) has been devised to aid analysis of materials handling, which is obviously less efficient when a material asset is handled any more than a minimally necessary number of times.

Types of Material Handling Equipment

Storage and Handling Equipment: Storage and handling equipment is a category within the material handling industry. The equipment that falls under this description is usually non-automated storage equipment. Products such as pallet racking, shelving and carts, among others, belong to storage and handling. Many of these products are often referred to as "catalogue" items because they generally have globally accepted standards and are often sold as stock materials out of Material handling catalogue.

Engineered Systems

Engineered systems are typically custom engineered material handling systems. Conveyors, Handling Robots, AS/RS, AGV and most other automated material handling systems fall into this category. Engineered systems are often a combination of products integrated to one system. Many distribution centres will optimize storage and picking by utilizing engineered systems such as pick modules and sortation systems.

Equipment and utensils used for processing or otherwise handling edible product or ingredients must be of such material and construction to facilitate thorough cleaning and to ensure that their use will not cause the adulteration of product during processing, handling, or storage. Equipment and utensils must be maintained in sanitary condition so as not to adulterate or contaminate product.

Industrial Trucks

Industrial trucks usually refer to operator driven motorized warehouse vehicles, powered manually, by gasoline, propane or electrically. Industrial trucks assist the material handling system with versatility; they can go where engineered systems cannot. Forklift trucks are the most common example of industrial trucks but certainly aren't the extent of the category. Tow tractors and stock chasers are additional examples of industrial trucks. Their greatest advantage lies in the wide range of attachments available; these increase the truck ability to handle various types and shapes of material.

Bulk Material Handling

Bulk material handling equipment is used to move and store bulk materials such as ore, liquids, and cereals. This equipment is often seen on farms, mines, shipyards and refineries. This category is also explained in Bulk material handling.

On-Rails Transfer Cart

On-rails transfer cart is a kind of material handling equipment. It moves on the rails and can transfer heavy cargoes or equipment with the weight 1-300t between the workshops or warehouses in the factory. It is widely used in the line of metallurgy, coal, heavy manufacturing, automotive assembly, etc. Its power can be AC or DC. DC Power has rail transmit power and battery power, while AC power includes cable power and slippery touch line power. In addition, there is the manual *rail transfer cart* or towed rail transfer cart, also called *motorized transfer trolley.*

Conveyors

Conveyors are another form of material handling. Conveyors can be used in a multitude of ways from warehouses to airport baggage handling systems. Some types of conveyors are unibilt, power and free, chain, towline and roller conveyor.

Cantilevered Crane Loading Platform

Cantilevered crane loading platforms are temporary platforms attached to the face of multi-storey buildings or structures to allow materials and equipment to be directly loaded on or shifted off floor levels by cranes during construction or demolition. They may be fixed or rolling and a variety of designs are used including fully fabricated and demountable types. The platforms are supported on needles (cantilevered beams) anchored to the supporting structure.

Energy for Agriculture

Energy has always been essential for the production of food. Prior to the industrial revolution, the primary energy input for agriculture was the sun; photosynthesis enabled plants to grow, and plants served as food for livestock, which provided fertilizer (manure) and muscle power for farming. However, as a result of the industrialization and consolidation of agriculture, food production has become increasingly dependent on energy derived from fossil fuels.

Today, industrial agriculture consumes fossil fuels for several purposes:

Industrial farms use huge quantities of synthetic fertilizers, which require significant energy inputs (primarily natural gas) to be produced. Other fertilizing agents (e.g., potassium and phosphorus) are mined, consuming even more energy.

Industrial agriculture is incredibly water intensive. The scarce resource is used for crop irrigation, which accounts for 31 percent of all water withdrawals in the US, waste management (i.e., for flushing manure out of industrial livestock facilities) and as drinking water for animals. This overuse of water has implications in the energy sector as well.

Modern agriculture relies upon machinery that runs on gasoline and diesel fuel (e.g., tractors and combines), and equipment that uses electricity (e.g., lights, pumps, fans, etc.).

Much of the food produced today is highly processed and heavily packaged, which further increases its energy footprint. As a result of consolidation and centralization of production, foods are often transported long distances, requiring additional energy inputs.

Most meat, eggs and dairy products are now produced on factory farms, huge industrial livestock operations that raise thousands of animals in confined conditions without access to pasture. Since the animals are unable to graze, factory farms require tremendous quantities of feed produced by industrial crop farms using the energy-intensive processes described above. Factory farms are also potential sources of ground and surface water pollution, which ultimately requires municipalities and private landowners to expend additional energy on water treatment.

Some factory farms use methane digesters to generate energy (digesters capture methane released during the decomposition of the huge quantities of manure generated onsite, and then burn the gas to produce electricity). Although this reduces emissions of methane (a potent greenhouse gas), the technology doesn't eliminate solid waste, fails to address other environmental, human health, social and animal welfare problems created by factory farms, and typically requires large subsidies to remain economically viable. Thus, despite being touted as a "green" energy source, methane digesters ultimately serve to subsidize and further entrench the environmentally and socially destructive model of industrial livestock production.

Energy policy also affects agriculture. For instance, congressional mandates now require the production of billions of gallons of ethanol, which is primarily—and controversially—derived from corn. Corn grown for ethanol takes land away from food production and, in states where corn is irrigated, uses a significant amount of water.

Given the growing population's food requirements, the world's finite supply of fossil fuels and the adverse environmental impacts of using this nonrenewable resource, the existing relationship between agriculture and energy must be dramatically altered. Among the most obvious solutions is to simply improve the energy efficiency of food production and distribution. This can be accomplished by shifting from energy-intensive industrial agricultural techniques to less intensive methods (e.g., pasture-raised livestock, drip irrigation, non-synthetic fertilizers, no-till crop management, etc.), using more efficient machinery and equipment, reducing food processing and packaging, promoting decentralization of food production and improving the efficiency of food transportation.

Farms can also generate their own clean electricity. While houses, barns and other buildings provide ample roof space for the installation of solar panels, farms with large swaths of land in windy areas are ideal sites for wind turbines. By leasing property for wind power production, these farms can earn an additional source of revenue while continuing to grow crops on surrounding land.

Despite the challenges posed by the energy-intensive nature of agriculture, the prudent use of resources and judicious application of technology has the capacity to significantly improve the long-term sustainability of food production.

Entry Levels for Interventions

It is useful to consider three entry levels for interventions as a means of examining both the energy needs for agriculture and the requirements for rural energy services in developing countries. These three levels are based on the "energy ladder" approach. For agriculture, the three-stage evolution can be considered as follows:

- basic human work for tilling, harvesting and processing, together with rain-fed irrigation, none of which involve an input from an external fuel source;
- then the use of animal work to provide various energy inputs;
- finally, the application of renewable energy technologies such as wind pumps, solar dryers and water wheels, together with modern renewable and fossil fuel based technologies for motive and stationary power applications, and for processing agricultural products.

For rural energy, the needs of poor people can be considered at the following three levels:

- energy for basic survival in cooking, lighting and space heating using traditional biomass fuels;
- then as people move up from subsistence, alternatives to traditional biomass fuels in these applications such as kerosene and LPG;
- finally, the role of enhanced energy services in rural areas using modern renewable energy and fossil fuels, for example in providing energy for small electrical appliances (e.g. lighting and radio) and the provision of community facilities (e.g. street lighting, water pumping, power for health centres and schools).

In both household and economic activities, the "energy ladder" follows and influences the "economic ladder". Attempts to alleviate poverty and to promote rural economic development and food security must be accompanied with efforts to promote the key role of energy, not simply as a goal in itself but as a vital component of these attempts.

The framework includes household, agriculture, small-scale rural industry and transport end-use requirements, and is based on empirical evidence. It illustrates the progression to modern fuels as income rises, and is based on empirical evidence gathered by the World Bank and other agencies. The data show that rural people choose to spend a significant proportion of their incomes on a better source of energy if they have access. The table shows how opportunities generally increase and the efficiency of energy utilization also increases, while the negative impacts of energy use decrease with rising household incomes.

It can be seen that woodfuel still plays an important role for households even at higher income levels. In agriculture and industry, diesel engines and electricity replace human and animal work; where rural electrification is not available or is too costly, diesel generators may be used instead. Wind pumping for water extraction from wells, together with mini-hydro and, more recently, PV systems for small-scale electricity supplies for homes, farms and community buildings are possible renewable energy options. The use of these options may appear in future empirical evidence.

Table: *Levels of household income and energy services*

End use	*Household income*		
	Low	*Medium*	*High*
Household			
Cooking	Wood, residues, dung	Wood, charcoal, dung, kerosene, biogas	Wood, charcoal, coal, kerosene, biogas, LPG, electricity
Lighting	Candles and kerosene	Candles, kerosene, gasoline	Kerosene, electricity, gasoline
Space heating	Wood, residues, dung	Wood, charcoal, dung	Wood, charcoal, dung, coal
Other appliances	Batteries (if any)	Electricity, batteries	Electricity, batteries
Agriculture			
Tilling	Human	Animal	Animal, gasoline, diesel
Irrigation	Human	Animal, wind pumps	Diesel, electricity
Post-harvest processing	Human, sun drying	Animal, water mills, sun drying	Diesel, electricity, solar drying
Rural Industry			
Mechanical tools	Human	Human, animal	Human, animal, diesel, electricity
Process heat	Wood, residues	Coal, charcoal, wood, residues	Coal, charcoal, wood, kerosene, residues
Transport			
Motive power	Human	Human, animal	Human, animal, diesel, gasoline

Energy and Agricultural Production

Agriculture is itself an energy conversion process, namely the conversion of solar energy through photosynthesis to food energy for humans and feed for animals. Primitive agriculture involved little more than scattering seeds on the land and accepting the scanty yields that resulted. Modern agriculture requires an energy input at all stages of agricultural production such as direct use of energy in farm machinery, water management, irrigation, cultivation and harvesting.

Post-harvest energy use includes energy for food processing, storage and in transport to markets. In addition, there are many indirect or sequestered energy inputs used in agriculture in the form of mineral fertilizers and chemical pesticides, insecticides and herbicides.

Whilst industrialized countries have benefited from these advances in energy availability for agriculture, developing countries have not been so fortunate. "Energizing" the food production chain has been an essential feature of agricultural development throughout recent history and is a prime factor in helping to achieve food security. Developing countries have lagged behind industrialized countries in modernizing their energy inputs to agriculture.

Agriculture accounts for only a relatively small proportion of total final energy demand in both industrialized and developing countries[8]. In OECD countries, for example, around 3-5% of total final energy consumption is used directly in the agriculture sector. In developing countries, estimates are more difficult to find, but the equivalent figure is likely to be similar - a range of 4-8% of total final commercial energy use.

The data for energy use in agriculture also exclude the energy required for food processing and transport by agro-industries. Estimates of these activities range up to twice the energy reported solely in agriculture. Definitive data do not exist for many of these stages, and this is a particular problem in analysing developing country energy statistics. In addition, the data conceal how effective these energy inputs are in improving agricultural productivity. It is the relationships between the amounts and quality of the direct energy inputs to agriculture and the resulting productive output that are of most interest.

Commercial Energy Use and Agricultural Output

On a broad regional basis, there appears to be a correlation between high per capita modern energy consumption and food production.

Whilst broad data on a regional basis conceal many differences between countries, crop types and urban and rural areas, the correlation is strong in developing countries, where higher inputs of modern energy can be assumed to have a positive impact on agricultural output and food production levels. The correlation is less strong in industrialized regions where food production is near or above required levels and changes in production levels may reflect changes in diet and food fashion rather than any advantages gained from an increased supply of modern energy.

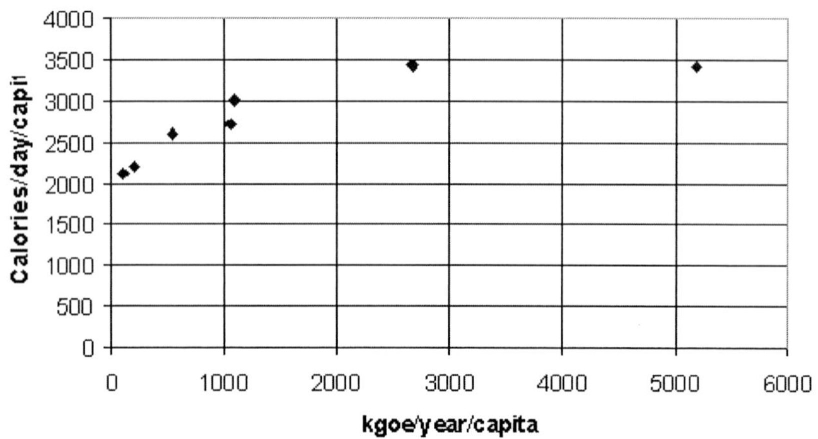

***Figure:** Modern energy consumption and food intake*

Empirical evidence suggests that the availability of modern energy such as petroleum fuels has proven to be essential in increasing the productivity of the agricultural sector in industrialized countries. In terms of the energy used per agricultural worker, the differences are even more dramatic, with developing countries using less than 5% of the energy per agricultural worker compared with industrialized countries. An obvious difference between industrialized and developing countries is the large numbers of agricultural workers per hectare in developing countries compared with industrialized countries.

In general, those regions with higher energy consumption have higher agricultural yields. However, the relationships between energy input and agricultural output are also affected by the varying ecological and environmental conditions around the world; soil fertility and rain-fed water availability being prime examples. Exact comparisons at the national level are, therefore, not easily made, but it is possible to use energy inputs for specific crops to gain further insights into the relationship between energy use and agricultural productivity.

As an example of this, a comparison between the commercial energy required for rice and maize production by modern methods in the United States, and transitional and traditional methods used in the Philippines and in Mexico is shown in the following table. These data show that the modern methods give greater productive yields and are much more energy-intensive than transitional and traditional methods. These methods include the use of fertilizer and other chemical inputs, more extensive irrigation and mechanized equipment.

Table: *Rice and maize production by modern, transitional and traditional methods*

	Rice production			***Maize production***	
	Modern (United States)	Transitional (Philippines)	Traditional (Philippines)	Modern (United States)	Traditional (Mexico)
Energy input (MJ/ha)	64,885	6,386	170	30.034	170
Productive yield (kg/ha)	5,800	2,700	1,250	5,083	950
Energy input yield (MJ/kg)	11.19	2.37	0.14	5.91	0.18

Agricultural Energy Needs

Agriculture practices in many developing countries continue to be based to a large extent on animal and human energy. Insufficient mechanical and electrical energy is available for agriculture, and hence the potential gains in agricultural productivity through the deployment of modern energy services are not being realized. Agricultural energy demand can be divided into direct and indirect energy needs. The direct energy needs include energy required for land preparation, cultivation, irrigation, harvesting, post-harvest processing, food production, storage and the transport of agricultural inputs and outputs. Indirect energy needs are in the form of sequestered energy in fertilizers, herbicides, pesticides, and insecticides.

Mankind has adapted a variety of resources to provide for energy in agriculture. Animal draught power obtained through domestication of cattle, horses and other animals has existed for over 8,000 years, the water wheel is over 2,000 years old, and windmills were introduced over 1,000 years ago. Direct sun energy for drying and biomass fuels for heating have also been prominent as agricultural energy inputs for centuries.

The bulk of direct energy inputs in developing countries, particularly in the subsistence agriculture sector, is in the form of human and animal work. Human work has a limited output, but humans are versatile, dextrous and can make judgements as they work. This gives humans an advantage in skilled operations such as transplanting, weeding, harvesting of fruits and vegetables and working with fibres. Water lifting and soil preparation need less skill but more energy input. A sustainable rate at which a fit person can use up energy is around 250-300 W, depending on climate and needing 10-30 minutes/hour rest. The efficiency of energy conversion is only about 25%, with a maximum sustainable power output of 75 W. Table 2.4 lists human power consumption for various farming activities.

Table: *Human power consumption for various farming activities*

Activity	***Gross power consumed (W)***
Clearing bush and scrub/felling trees	400-600
Hoeing, planting	200-500
Ridging, deep digging	400-1000
Ploughing with draught animal	350-550
Driving 4 wheel tractor	150-300
Driving single axle tractor	350-650

Animals provide transport of products, can pull implements and lift water and are used in processing activities such as cane crushing and threshing. The world population of draught animals is estimated at over 400 million. Animal work can alleviate human drudgery and increase agricultural production. Power output ranges from 200 W for a donkey to over 500 W for a buffalo, and daily working hours range from 4 hours for a donkey up to 10 hours for a horse. There are limitations on the performance of animals, especially at times when they are needed most in the dry season with feed, grazing and water in short supply. Implements used by animals can be pole-pull, chain-pull or wheeled carriers.

Table: *Power and energy output of individual farm animals*

Animal	***Typical weight(kN)***	***Pull-weight ratio***	***Power output (W)***	***Energy output per day (MJ)***
Ox	4.5	0.11	450	10
Buffalo	5.5	0.12	520	9.5
Horse	4.0	0.13	500	18
Donkey	1.5	0.13	200	3
Mule	3.0	0.13	400	8.5
Camel	5.0	0.13	650	14

It seems likely that both human and animal work will continue to be used as agricultural inputs for the foreseeable future in developing countries. Efforts to support these farming traditions include work on animal efficiency, which can be improved through modernization of equipment, better breeding and animal husbandry, feeding and veterinary care, and on improved designs of animal-drawn farm equipment.

Mechanization and Conservation Agriculture

A transition to increased mechanization of farm operations has been undertaken over a long period of time. Tillage operations are used within arable farming systems, which are often the operations with the highest energy requirements. Mechanized soil tillage has in the past been associated with increased soil fertility, due to the mineralization of soil nutrients; it also allows higher working depths and speeds and the use of implements such as ploughs, disc harrows and rotary cultivators. This process leads in the long term to a reduction of soil organic matter, and most soils degrade under long lasting intensive arable agriculture with particularly detrimental effects on soil structure. The process is dramatic under tropical climate conditions but can be noticed all over the world. Reduction of mechanical tillage and promotion of soil organic matter through permanent soil cover is an approach to reverting soil degradation and other environmental impacts of conventional agriculture, achieving at the same time a high agricultural production level on truly sustainable basis. This approach is described as 'conservation agriculture' and replaces mechanical soil tillage by 'biological tillage'.

In this approach, crop residues remaining on the soil surface produce a layer of mulch which protects the soil from the physical impact of rain and wind, and also stabilizes the soil surface layers moisture content and temperature. This zone becomes a habitat for a number of organisms which macerate the mulch, mix it with soil and assist its decomposition to humus. Agriculture with reduced mechanical tillage is only possible when soil organisms take over the task of tilling the soil. This leads to other implications regarding the use of chemical inputs since synthetic pesticides and mineral fertilizers have to be used in a way that does not harm soil life. Conservation agriculture can only work if all agronomic factors are equally well managed.

It should also be noted that, whilst shifts from human work to mechanized processes may offer more efficient use of resources, and deliver productive and economic benefits, the agricultural sector does provide a source of paid employment for rural people in developing countries. A balance is often required between the socio-economic benefits and the agricultural productive benefits of changing the processes (and thereby the energy inputs) involved.

Chemical Inputs

Fertilizers, and other chemical inputs to agriculture, have proved important in the past in increasing food production in all regions of the world. Mineral fertilizers, chemical pesticides, fungicides and herbicides all require energy in their production, distribution and transport processes. Fertilizers form the largest of these energy inputs to agriculture, whilst pesticides are the most energy-intensive agricultural input (on a per kg basis of chemical). The need for insecticides and fungicides can, however, be reduced through greater use of pest control methods based on the principals of integrated pest management.

The energy contents of various agricultural inputs, which show illustrative values of the energy required for manufacturing these products. Full life-cycle values (i.e. from extracting the chemical raw material through to final delivery of the manufactured product to the field) would depend on several highly variable factors. These include the actual location of chemical raw material supplies relative to the productive facility, the ease of extraction, the chemical manufacturing processes, and the distances involved in transporting raw, semi-finished and finished products.

Table: *Energy content of agricultural inputs*

Input		*Typical rate of application (kg/ha)*	*Sequestered energy (MJ/kg)*	*Energy content of crop produce (MJ/ha)*
Fertilizer	Nitrogen	150	65	9,750
	Phosphate	60	9	540
	Potash	60	6	360
Insecticide		0.14	200	28
Herbicide		5	240	1,200
Fungicide		3	92	276
Seed		120	14	1,680

By comparison with the energy content of mineral fertilizer, the energy content of fresh manure is much lower at around 0.35 MJ/kg. However, the amount of plant, human and animal wastes available in developing countries that could potentially be used for organic manure is several times the consumption of chemical fertilizers in these countries. The volume of fertilizer input to agriculture differs markedly between regions of the world. The data, which are broad regional estimates, show that Africa and Latin America use very low inputs compared with Asia and Europe.

Table: *Fertilizer input to agriculture*

	Average annual fertilizer use (kg/ha crop land)
World	96
Africa	20
Asia	123
Europe	192
North and Central America	87
Latin America	44

Agricultural practices are, however, changing throughout the world. In recent years, diminishing returns from increased use of fertilizers and pesticides have raised questions about their role in the transition towards more environmentally-sound agricultural practices. The main constraints in increasing organic manure use are the large labour and skills content required together with the need to change cultural attitudes and the lack of mixed livestock and crop husbandry.

Cooking

Energy is required for cooking most foods, and for boiling water for drinking, washing and for partial water purification. In small-scale food cooking, many rural households continue to rely on woodfuel. The data are derived from a combination of the energy content of each fuel and the efficiency with which the fuels are typically burned for cooking in developing countries.

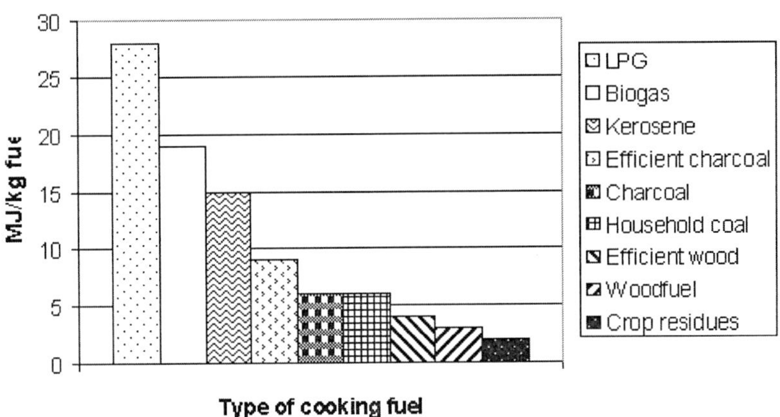

Figure: *Energy efficiency of selected cooking fuels*

Traditional biomass are generally much less efficient for cooking than modern, commercially traded fuels such as liquefied petroleum gas and kerosene. These fuels have higher energy values per unit of weight than woodfuels, and are generally used in more efficient stoves. In addition, the level of heat output of kerosene stoves can be adjusted, so making kerosene more convenient for preparing a wide range of foods. Another versatile fuel is biogas, derived from digesters using dung and farm residues, and both China and India have done much to develop biogas and encourage its use among people in rural areas. The least efficient fuels are agricultural residues, but poor people are forced to use these fuels where they are available in the local environment because they have no cash cost. Improved cooking stove programmes have been a feature of many rural energy interventions over the last 20 years. Work has been carried out on establishing multi-sectoral support and involvement, and integrating technology, dissemination and financing. National standards on cooking and heating efficiency and guidelines for stove programmes have been developed. Studies have shown that the health and quality of life of both women and children can be improved by improving access to biomass and providing biomass stoves which are designed to be safer to use in terms of reduced risk of burns, respiratory disease and eye problems. Indeed, the thrust of the current work on improved cooking stoves has become one of reducing the adverse health impact of traditional cooking methods rather than one of achieving potential fuel savings or increasing stove energy efficiency.

Energy and Agroprocessing

The agroprocessing industry transforms products originating from agriculture into both food and non-food commodities. Processes range from simple preservation (such as sun drying) and operations closely related to harvesting, to the production, by modern, capital-intensive methods of such articles as textiles, pulp and paper. Upstream industries are engaged in the initial processing of products, with examples such as rice and flour milling, leather tanning, cotton ginning, oil pressing, saw milling and fish canning. Downstream industries undertake further manufacturing operations on intermediate products made from agricultural materials. Examples are bread and noodle making, textile spinning and weaving, paper production, clothing and footwear manufacture and rubber manufacture.

An energy input is required in food processing, as well as in packaging, distribution and storage. Many food crops when harvested cannot be consumed directly, but must pass through several stages of processing as well as cooking in order to be palatable and digestible. Raw meats, uncooked grains, vegetables and even fruits require preparation and heating to enhance their flavour, rendering their components edible and digestible. The processing and cooking stages reduce harmful organisms and parasites, which might pose health hazards.

Poorly handled and stored food can become spoiled and contaminated. Food preservation usually requires the application of heat to destroy micro-biological agents, such as bacteria, yeast and mould. Pasteurization causes the inactivation of spoilage enzymes and reduction of bacteria at temperatures around 80-90°C. Heat sterilization can use atmospheric steam at 100°C for high-acid foods, and pressurized steam at around 120°C for low acid foods. Other techniques include dehydration to reduce moisture content, pickling/ smoking to reduce microbial activity, fermentation, salting and freezing.

Food transformation activities are generally less energy-intensive and release less CO_2 and metal residues than most other industrial activities per unit of product. As described in more detail below, agroprocessing industries, such as sugar mills, can become not only energy self-sufficient through the energy conversion of biomass residues, but also electricity producers for export to other users.

Energy Policies in Agriculture

The international community has agreed that a priority in Agenda 21 should be promoting sustainable agriculture and rural development (SARD) and one of the programme areas. The objectives of this programme area include initiating the required transition by making available new and renewable sources of energy and increasing the energy inputs available for rural household and agro-industrial needs. Rural programmes favouring sustainable development of renewable energy sources and improved energy efficiency are also called for in Agenda 21 in the programme area dealing with protection of the atmosphere.

The international framework for action does, therefore, exist. Nevertheless, it seems clear from the data presented that there is a wide disparity between energy inputs in agriculture in industrialized countries with that in developing countries and that this disparity is hindering agricultural productivity gains. One consequence of this is that enhanced food security is more difficult to achieve. This affects not only the quantity of food produced, but also the quality - people are forced to eat uncooked food or food that can easily be cooked but which may not give full nourishment.

This situation is manifest in the general lack of an agricultural component in rural energy development policies. Agriculture contributes significantly to economic and social development, accounting for around 30% of developing country GDP, but energy provision for agriculture has not received the attention that the sector deserves. The World Bank has observed (World Bank, 2000b) that most energy policies in developing countries have traditionally focused on large capital investments in the generation and transmission of electricity, gas and petroleum products, so enabling the commercial development of energy supply industries. These policies are designed mainly for the needs of industry, transport and urban infrastructure, and thereby focus most attention on urban populations, whilst rural populations (and to some extent peri-urban populations) and their energy requirements are frequently overlooked.

Even where there has been a focus on rural energy and development and attention given to the scope for improving energy services to rural populations, much of the work has concentrated on the household use of energy. For example, there have been many programmes to expand the use of improved cooking stoves and to provide households and community buildings with small amounts of electrical power through the development of solar home systems using PV technology. Whilst these PV programmes offer many benefits at the household level, none of them have directly tackled the energy supply problems in the agriculture sector, and nor have they promoted income generating activities. Such activities are the only way for rural people to break out from the vicious circle of poverty.

FAO has been active in assisting developing countries to meet their energy requirements in agriculture, forestry and fisheries as a means of achieving sustainable rural development. An integrated approach to incorporating energy into rural and agricultural planning has been promoted, together with the increased use of modern energy technologies. Activities sponsored by FAO have included, for example, a project in China, which integrated alcohol production from sorghum with biogas, pyrolysis, solar and wind energy systems. This work also examined energy conservation and the potential of various renewable energy sources in specific farm activities. The project is presently being expanded to produce ethyl tertiary butyl ether (ETBE) from ethanol. ETBE is then mixed with gasoline and offers reduced air pollution and CO_2 emissions from transport. Trials are planned for Shanghai and Shenyang.

All agricultural production has to become commercial to enable long-term food security; as one of the prime inputs to agriculture, energy plays a decisive role in attaining food security. Attempts to alleviate hunger and to promote rural development and food security must be accompanied by efforts to promote the key role of energy, not as a goal in itself but as a vital component of these attempts. In many cases, it seems that the low quality and the meagre amounts of energy available for the food production and supply chain are at the heart of food security problems.

The empirical evidence and examples presented earlier indicate that it is possible to quantify the extent of this gap in energy terms. However, if this gap were to be closed by an over-zealous use of fossil fuels, then some of the adverse environmental and resource utilization issues identified will become more pressing. This solution would not offer a sustainable means of tackling the energy gap problem. Development and deployment of renewable energy systems together with improved end-use energy efficiency techniques can provide an alternative means of helping to bridge this gap. A large number of such energy technologies are mature and commercially available, whilst others still require further research or demonstration. Actions to invest in cost-effective systems and to develop the most promising new technologies are needed.

Impact of Trade on Energy Demand in Agriculture

The recent moves towards opening world markets to a much wider agricultural trade will have an impact on energy requirements in the agricultural sector. Greater international trade will encourage and require more energy for processing primary agricultural products closer to the source of production in developing countries, then transporting, storing and finally distributing these products to industrialized countries. For example, horticultural products often have a high added value and a transportation and storage system between the farm and the end market using a cold-chain or a controlled modified atmosphere is needed to maintain this value. Hence the new trade opportunities point to additional energy intensive applications developing in the agricultural sector, so increasing the level of energy demand in developing countries.

A New Direction for Energy in Agriculture - the Use of Measurable Indicators

Ways of measuring the sustainability of energy and agriculture and in particular the benefits of improved energy services for agriculture need to be established, so that indicators of progress can be developed. For example, current work on sustainability indicators could be adapted to include energy and agriculture. Such measures would assist the establishment of a bridge between the rural energy and agricultural policy dimensions so that agricultural energy needs can be included in overall energy planning and policy formation. FAO has initiated work in this field and is developing energy indicators of sustainable agriculture. A third linkage is between the environmental dimension and energy use by agriculture, and suitable indicators of this also need to be developed.

Energy problems and solutions for agriculture should always be guided by local economic, environmental and social considerations. Energy policy formation should bring together national energy development policies with the locally perceived priorities. There needs to be increased emphasis on non-fossil fuel alternatives to the provision of energy services in agriculture in developing countries (Best, 1997). These range from the modern renewable energy sources, such as improved biomass conversion (including liquid biofuels, biogas, gasification), solar energy (PV), wind and geothermal energy and small-scale hydropower, to lower energy intensity industries, material and energy recycling and better means of utilizing traditional energy sources, such as improved cooking stoves. In addition, there needs to be improved energy efficiency in mechanical equipment and in drying and separating operations.

The agriculture sector can move towards a path of greater sustainability through the application of improved techniques and practices such as conservation agriculture, organic farming, protecting agro-biodiversity, better water and soil management, and integrated pest management and plant nutrition. Where appropriate a greater level of mechanization and improved food-processing technologies are also important. The main challenge in the medium term for energy in agriculture is to mobilize the changes occurring in both the energy supply and the agricultural sectors to the benefit of rural livelihoods and communities. There is a danger that rural populations could be left behind unless energy policies are directed specifically on their needs. It seems clear from the analysis presented that energy requirements should be appropriately considered and integrated into agricultural and rural development programmes.

Renewable Energy Management

Renewable energy is a socially and politically defined category of energy sources. Renewable energy is generally defined as energy that comes from resources which are continually replenished on a human timescale such as sunlight, wind, rain, tides, waves and geothermal heat.

About 16% of global final energy consumption comes from renewable resources, with 10% of all energy from traditional biomass, mainly used for heating, and 3.4% from hydroelectricity. New renewables (small hydro, modern biomass, wind, solar, geothermal, and biofuels) accounted for another 3% and are growing rapidly. The share of renewables in electricity generation is around 19%, with 16%

of electricity coming from hydroelectricity and 3% from new renewables. While many renewable energy projects are large-scale, renewable technologies are also suited to rural and remote areas, where energy is often crucial in human development.

Renewable energy sources, that derive their energy from the sun, either directly or indirectly, such as Hydro and wind, are expected to be capable of supplying humanity energy for almost another 1 billion years, at which point the predicted increase in heat from the sun is expected to make the surface of the Earth too hot for liquid water to exist.

Worldwide, agriculture contributes between 14 and 30 percent of human-caused greenhouse gas (GHG) emissions because of its heavy land, water, and energy use—that's more than every car, train, and plane in the global transportation sector. Livestock production alone contributes around 18 percent of global emissions, including 9 percent of carbon dioxide, 35 percent of methane, and 65 percent of nitrous oxide.

Activities like running fuel-powered farm equipment, pumping water for irrigation, raising dense populations of livestock in indoor facilities, and applying nitrogen-rich fertilizers all contribute to agriculture's high GHG footprint.

The good news? The UN Food and Agriculture Organization (FAO) estimates that the sector has "significant" potential to reduce its emissions, including removing 80 to 88 percent of the carbon dioxide that it currently produces.

Some of this reduction can be achieved by substituting renewable energy for the fossil fuels typically used to power day-to-day farm activities. Do-it-yourself solar heat collectors can warm livestock buildings, greenhouses, and homes; small or cooperatively owned wind and water turbines can pump water and power equipment; photovoltaic panels can power critical farm operations like electric fencing and drip irrigation systems; and designing or renovating buildings and barns to maximize natural daylight can dramatically reduce the electricity required to light and warm farm buildings.

These innovations can be scaled up for implementation on large farms, but their beauty is in their simplicity, accessibility, and application to the smallest of operations.

And greening a farm does not stop at replacing fossil fuels with renewable energy. To make a farm truly climate-smart, it must take into account all aspects of its environmental footprint: soil fertility, water use, chemical inputs, and biodiversity. Farmers can implement low-tech, low-cost practices to curb their emissions while building resilience to weather shocks and severe resource scarcity, two projected stumbling blocks for farmers in coming decades.

Any measure that reduces on-farm water use, for instance, will help to relieve the heavy pressures on the planet's dwindling resources while reducing agriculture's energy footprint. Agriculture accounts for a whopping 70 percent of global water use; in the United States, the figure rises to 80 percent. Rivers, lakes, and underwater aquifers are drying up across the globe, causing serious concerns over basic human rights like sanitation, food production, and safe drinking water.

Installing drip irrigation, which applies precise amounts of water to the plant roots instead of spraying water over plants, is a simple way to invest in climate-stress mitigation. Watering crops using "greywater," or water used in domestic activities like dishwashing, laundry, and bathing — not to be confused with blackwater, or sewage — can also reduce water use on farms, particularly small-scale operations. And switching from "thirsty" crops like rice, wheat, and sugarcane (which account for nearly 60 percent of the world's irrigated cropland) to less-demanding plants like sorghum, millet, lettuce, broccoli, carrots, beans, and squash can reduce on-farm water use and help farmers cope with drought and other threats (while broadening access to fresh and nutritious foods).

Water conservation is just one approach to making a farm more climate-friendly. Practices such as using animal manure rather than artificial fertilizer, planting trees on farms to reduce soil erosion and sequester carbon, and growing food in cities all hold huge potential for reducing agriculture's environmental footprint.

Interestingly, biofuels can combine the need for renewable energy with climate-friendly agricultural practices.

Perennial bioenergy crops stand as a shining example of agricultural innovation. While corn has long reigned the biofuels industry, its relative energy-conversion inefficiency and its sensitivity to high temperatures make it an unsustainable long-term energy option. But trees and shrubs like willow, sycamore, sweetgum, and cottonwood offer promising alternatives. These perennials grow quickly for many years, can often thrive on marginal land, and are often much hardier than annual plants like corn or soybeans. Their long roots can also reduce erosion, filter groundwater, harbor beneficial microorganisms, and help soil retain key nutrients like phosphorus and nitrogen. By tapping into the multitude of climate-friendly farming practices that already exist, agriculture can continue to supply food for the world's population, and also help to reduce our dependence on fossil fuels. But if we want agriculture to contribute to climate change mitigation, climate-friendly food production will need to receive increased attention — in the form of both research and investment — in the coming years.

Biogas and Gasification

Biogas typically refers to a gas produced by the breakdown of organic matter in the absence of oxygen. It is a renewable energy source, like solar and wind energy. Furthermore, biogas can be produced from regionally available raw materials such as recycled waste and is environmentally friendly.

Biogas is produced by anaerobic digestion with anaerobic bacteria or fermentation of biodegradable materials such as manure, sewage, municipal waste, green waste, plant material, and crops. Biogas comprises primarily of methane (CH4) and carbon dioxide (CO_2) and may have small amounts of hydrogen sulphide (H2S), moisture and siloxanes.

The gases methane, hydrogen, and carbon monoxide (CO) can be combusted or oxidized with oxygen. This energy release allows biogas to be used as a fuel. Biogas can be used as a fuel in any country for any heating purpose, such as cooking. It can also be used in a gas engine to convert the energy in the gas into electricity and heat.

Biogas can be compressed, the same way natural gas is compressed to CNG, and used to power motor vehicles. In the UK, for example, biogas is estimated to have the potential to replace around 17% of vehicle fuel. Biogas is a renewable fuel so it qualifies for renewable energy subsidies in some parts of the world. Biogas can also be cleaned and upgraded to natural gas standards when it becomes bio methane.

Production

Bio gas is practically produced as landfill gas (LFG) or digested gas. A *bio gas plant* is the name often given to an anaerobic digester that treats farm wastes or energy crops. Bio gas can be produced using anaerobic digesters. These plants can be fed with energy crops such as maize silage or biodegradable wastes including sewage sludge and food waste. During the process, an air-tight tank transforms biomass waste into methane producing renewable energy that can be used for heating, electricity, and many other operations that use any variation of an internal combustion engine, such as GE Jenbacher gas engines.

Figure: *Biogas production in rural Germany*

There are two key processes: Mesophilic and Thermophilic digestion. In experimental work at University of Alaska Fairbanks, a 1000-litre digester using psychrophiles harvested from "mud from a frozen lake in Alaska" has produced 200–300 liters of methane per day, about 20–30% of the output from digesters in warmer climates. Landfill gas is produced by wet organic waste decomposing under anaerobic conditions in a landfill.

The waste is covered and mechanically compressed by the weight of the material that is deposited from above. This material prevents oxygen exposure thus allowing anaerobic microbes to thrive. This gas builds up and is slowly released into the atmosphere if the landfill site has not been engineered to capture the gas. Landfill gas is hazardous for three key reasons. Landfill gas becomes explosive when it escapes from the landfill and mixes with oxygen. The lower explosive limit is 5% methane and the upper explosive limit is 15% methane.

The methane contained within biogas is 20 times more potent a greenhouse gas than carbon dioxide. Therefore, uncontained landfill gas, which escapes into the atmosphere may significantly contribute to the effects of global warming. In addition, volatile organic compounds (VOCs) contained within landfill gas contribute to the formation of photochemical smog.

Composition

Table: Typical composition of biogas

Compound	Molecular formula	%
Methane	CH_4	50–75
Carbon dioxide	CO_2	25–50
Nitrogen	N_2	0–10
Hydrogen	H_2	0–1
Hydrogen sulphide	H_2S	0–3
Oxygen	O_2	0–0

The composition of biogas varies depending upon the origin of the anaerobic digestion process. Landfill gas typically has methane concentrations around 50%. Advanced waste treatment technologies can produce biogas with 55–75% methane, which for reactors with free liquids can be increased to 80-90% methane using in-situ gas purification techniques As-produced, biogas also contains water vapor. The fractional volume of water vapor is a function of biogas temperature; correction of measured gas volume for both water vapor content and thermal expansion is easily done via simple mathematics which yields the standardized volume of dry biogas.

In some cases, biogas contains siloxanes. These siloxanes are formed from the anaerobic decomposition of materials commonly found in soaps and detergents. During combustion of biogas containing siloxanes, silicon is released and can combine with free oxygen or various other elements in the combustion gas. Deposits are formed containing mostly silica (SiO_2) or silicates (Si_xO_y) and can also contain calcium, sulphur, zinc, phosphorus. Such white mineral deposits accumulate to a surface thickness of several millimeters and must be removed by chemical or mechanical means.

Practical and cost-effective technologies to remove siloxanes and other biogas contaminants are currently available.

Benefits

When biogas is used, many advantages arise. In North America, utilization of biogas would generate enough electricity to meet up to three percent of the continent's electricity expenditure. In addition, biogas could potentially help reduce global climate change. Normally, manure that is left to decompose releases two main gases that cause global climate change: nitrogen dioxide and methane. Nitrogen dioxide (NO_2) warms the atmosphere 310 times more than carbon dioxide and methane 21 times more than carbon dioxide. By converting cow manure into methane biogas via anaerobic digestion, the millions of cows in the United States would be able to produce one hundred billion kilowatt hours of electricity, enough to power millions of homes across the United States. In fact, one cow can produce enough manure in one day to generate three kilowatt hours of electricity; only 2.4 kilowatt hours of electricity are needed to power a single one hundred watt light bulb for one day. Furthermore, by converting cow manure into methane biogas instead of letting it decompose, global warming gases could be reduced by ninety-nine million metric tons or four percent. In Nepal biogas is being used as a reliable source of rural energy, says Bikash Haddi of Biogas promotion centre.

Applications

Biogas can be utilized for electricity production on sewage works, in a CHP gas engine, where the waste heat from the engine is conveniently used for heating the digester; cooking; space heating; water heating; and process heating. If compressed, it can replace compressed natural gas for use in vehicles, where it can fuel an internal combustion engine or fuel cells and is a much more effective displacer of carbon dioxide than the normal use in on-site CHP plants.

Methane within biogas can be concentrated via a biogas upgrader to the same standards as fossil natural gas, which itself has had to go through a cleaning process, and becomes *biomethane.* If the local gas network allows for this, the producer of the biogas may utilize the local gas distribution networks. Gas must be very clean to reach pipeline quality, and must be of the correct composition for the local distribution network to accept. Carbon dioxide, water, hydrogen sulfide, and particulates must be removed if present.

Biogas Upgrading

Raw biogas produced from digestion is roughly 60% methane and 29% CO2 with trace elements of H2S, and is not high quality enough to be used as fuel gas for machinery. The corrosive nature of H2S alone is enough to destroy the internals of a plant. The solution is the use of biogas upgrading or purification processes whereby contaminants in the raw biogas stream are absorbed or scrubbed, leaving more methane per unit volume of gas. There are four main methods of biogas upgrading, these include water washing, pressure swing absorption, selexol absorption, and amine gas treating.

The most prevalent method is water washing where high pressure gas flows into a column where the carbon dioxide and other trace elements are scrubbed by cascading water running counter-flow to the gas. This arrangement could deliver 98% methane with manufacturers guaranteeing maximum 2% methane loss in the system. It takes roughly between 3-6% of the total energy output in gas to run a biogas upgrading system....

Biogas Gas-grid Injection

Gas-grid injection is the injection of biogas into the methane grid (natural gas grid). Injections includes biogas: until the breakthrough of micro combined heat and power two-thirds of all the energy produced by biogas power plants was lost (the heat), using the grid to transport the gas to customers, the electricity and the heat can be used for on-site generation resulting in a reduction of losses in the transportation of energy. Typical energy losses in natural gas transmission systems range from 1–2%. The current energy losses on a large electrical system range from 5–8%.

Biogas in Transport

If concentrated and compressed, it can also be used in vehicle transportation. Compressed biogas is becoming widely used in Sweden, Switzerland, and Germany. A biogas-powered train, named *Biogaståget Amanda*, has been in service in Sweden since 2005. Biogas also powers automobiles and in 1974, a British documentary film entitled *Sweet as a Nut* detailed the biogas production process from pig manure, and how the biogas fueled a custom-adapted combustion engine. In 2007, an estimated 12,000 vehicles were being fueled with upgraded biogas worldwide, mostly in Europe.

Legislation

The European Union presently has some of the strictest legislation regarding waste management and landfill sites called the Landfill Directive. The United States legislates against landfill gas as it contains VOCs. The United States Clean Air Act and Title 40 of the Code of Federal Regulations (CFR) requires landfill owners to estimate the quantity of non-methane organic compounds (NMOCs) emitted. If the estimated NMOC emissions exceeds 50 tonnes per year, the landfill owner is required to collect the landfill gas and treat it to remove the entrained NMOCs. Treatment of the landfill gas is usually by combustion. Because of the remoteness of landfill sites, it is sometimes not economically feasible to produce electricity from the gas. However, countries such as the United Kingdom and Germany now have legislation in force that provides farmers with long-term revenue and energy security.

Development Around the World

United States: With the many benefits of biogas, it is starting to become a popular source of energy and is starting to be utilized in the United States more. In 2003, the United States consumed 147 trillion BTU of energy from "landfill gas", about 0.6% of the total U.S. natural gas consumption. Methane biogas derived from cow manure is also being tested in the U.S. According to a 2008 study, collected by the *Science and Children* magazine, methane biogas from cow manure would be sufficient to produce 100 billion kilowatt hours enough to power millions of homes across America. Furthermore, methane biogas has been tested to prove that it can reduce 99 million metric tons of greenhouse gas emissions or about 4% of the greenhouse gases produced by the United States.

In Vermont, for example, biogas generated on dairy farms around the state is included in the CVPS Cow Power programme. The Cow Power programme is offered by Central Vermont Public Service Corporation as a voluntary tariff. Customers can elect to pay a premium on their electric bill, and that premium is passed directly to the farms in the programme. In Sheldon, Vermont, Green Mountain Dairy has provided renewable energy as part of the Cow Power programme. It all started when the brothers who own the farm, Bill and Brian Rowell, wanted to address some of the manure management challenges faced by dairy farms, including manure odor, and nutrient availability for the crops they need to grow to feed the animals. They installed an anaerobic digester to process the cow and milking centre waste from their nine hundred and fifty cows to produce renewable energy, a bedding to replace sawdust, and a plant friendly fertilizer. The energy and environmental attributes are sold. On average, the system run by the Rowell brothers produces enough electricity to power three hundred to three hundred fifty other homes. The generator capacity is about 300 kilowatts.

In Hereford, Texas, cow manure is being used to power an ethanol power plant. By switching to methane biogas, the ethanol power plant has saved one thousand barrels of oil a day. Overall, the power plant has reduced transportation costs and will be opening many more jobs for future power plants that will be relying on biogas.

Europe

The level of development varies greatly in Europe. While countries such as Germany, Austria and Sweden are fairly advanced in their usage of biogas, there is still a vast potential for this renewable energy source in the rest of the continent, especially in Eastern Europe. Different legal frameworks, education schemes and the availability of technology are among the prime reasons behind this untapped potential. Another challenge for the further progression of biogas has been negative public perception.

Initiated by the events of the gas crisis within Europe during December 2008, it was decided to launch the EU project "SEBE" (Sustainable and Innovative European Biogas Environment) which is financed under the CENTRAL programme. The goal is to address the energy dependence of Europe by establishing an online platform to combine available knowledge and launch pilot projects aimed at raising awareness among the public and developing new biogas technologies.

In February 2009, the European Biogas Association (EBA) was founded in Brussels as a non-profit organisation to promote the deployment of sustainable biogas production and use in Europe. EBA's strategy defines three priorities: establish biogas as an important part of Europe's energy mix, promote source separation of household waste to increase the gas potential and support the production of biomethane as vehicle fuel. In July 2013, it had 60 members from 24 countries across Europe.

UK

There are currently around 60 non-sewage biogas plants in the UK, most are on-farm, but some larger facilities exist off-farm, which are taking food and consumer wastes.

On 5 October 2010, biogas was injected into the UK gas grid for the first time. Sewage from over 30,000 Oxfordshire homes is sent to Didcot sewage treatment works, where it is treated in an anaerobic digestor to produce biogas, which is then cleaned to provide gas for approximately 200 homes.

Germany

Germany is Europe's biggest biogas producer as it is the market leader in biogas technology. In 2010 there were 5,905 biogas plants operating throughout the whole country, in which Lower Saxony, Bavaria and the eastern federal states are the main regions. Most of these plants are employed as power plants. Usually the biogas plants are directly connected with a CHP which produces electric power by burning the bio methane. The electrical power is then fed into the public power grid. In 2010, the total installed electrical capacity of these power plants was 2,291 MW. The electricity supply was approximately 12.8 TWh, which is 12.6 per cent of the total generated renewable electricity. Biogas in Germany is primarily extracted by the co-fermentation of energy crops (called 'NawaRo', an abbreviation of 'nachwachsende Rohstoffe', which is German for renewable resources) mixed with manure, the main crop utilized is corn. Organic waste and industrial and agricultural residues such as waste from the food industry are also used for biogas generation. In this respect, Biogas production in Germany differs significantly from the UK, where biogas generated from landfill sites is most common.

Biogas production in Germany has developed rapidly over the last 20 years. The main reason for this development is the legally created frameworks. Governmental support of renewable energies started at the beginning of the 1990s with the Law on Electricity Feed (StrEG). This law guaranteed the producers of energy from renewable sources the feed into the public power grid, thus the power companies were forced to take all produced energy from independent private producers of green energy. In 2002 the Law on Electricity Feed was replaced by the Renewable Energy Source Act (EEG). This law even guaranteed a fixed compensation for the produced electric power over 20 years. The amount of ca. 0.08 Euro gave particular farmers the opportunity to become an energy supplier and gaining a further source of income in the same place.

The German agricultural biogas production was given a further push in 2004 by implementing the so-called NawaRo-Bonus. This is a special bonus payment given for the usage of renewable resources i.e. energy crops. In 2007 the German government stressed its intention to invest further effort and support in improving the renewable energy supply to provide an answer on growing climate challenges and increasing oil prices by the 'Integrated Climate and Energy Programme'.

This continual trend of renewable energy promotion induces a number of challenges facing the management and organisation of renewable energy supply that has also several impacts on the biogas production. The first challenge to be noticed is the high area-consuming of the biogas electric power supply. In 2011 energy crops for biogas production consumed an area of circa 800,000 ha in Germany. This high demand of agricultural areas generates new competitions with the food industries that did not exist yet. Moreover new industries and markets were created in predominately rural regions entailing different new players with an economic, political and civil background. Their influence and acting has to be governed to gain all advantages this new source of energy is offering. Finally biogas will furthermore play an important role in the German renewable energy supply if good governance is focused.

Indian Subcontinent

In India, Nepal, Pakistan and Bangladesh biogas produced from the anaerobic digestion of manure in small-scale digestion facilities is called gobar gas; it is estimated that such facilities exist in over two million households in India, fifty thousands in Bangladesh and thousands in Pakistan, particularly North Punjab, due to the thriving population of livestock. The digester is an airtight circular pit made of concrete with a pipe connection. The manure is directed to the pit, usually directly from the cattle shed. The pit is then filled with a required quantity of wastewater. The gas pipe is connected to the kitchen fireplace through control valves. The combustion of this biogas has very little odour or smoke. Owing to simplicity in implementation and use of cheap raw materials in villages, it is one of the most environmentally sound energy sources for rural needs. One type of these system is the Sintex Digester. Some designs use vermiculture to further enhance the slurry produced by the biogas plant for use as compost.

In order to create awareness and associate the people interested in biogas, an association "Indian Biogas Association" was formed. The "Indian Biogas Association" aspires to be a unique blend of; nationwide operators, manufacturers and planners of biogas plants, and representatives from science and research. The association was founded in 2010 and is now ready to start mushrooming. The sole motto of the association is "propagating Biogas in a sustainable way".

The Deenabandhu Model is a new biogas-production model popular in India. (*Deenabandhu* means "friend of the helpless.") The unit usually has a capacity of 2 to 3 cubic metres. It is constructed using bricks or by a ferrocement mixture. In India, the brick model costs slightly more than the ferrocement model; however, India's Ministry of New and Renewable Energy offers some subsidy per model constructed.

In Pakistan, the Rural Support Programmes Network is running the Pakistan Domestic Biogas Programme which has installed over 1500 biogas plants and has trained in excess of 200 masons on the technology and aims to develop the Biogas Sector in Pakistan.

Also PAK-Energy Solution has taken the most innovative and responsible initiatives in biogas technology. In this regard, the company is also awarded by 1st prize in "Young Entrepreneur Business Plan

Challenge" jointly organized by Punjab Govt. & LCCI. They have designed and developed Uetians Hybrid Model, in which they have combined fixed dome and floating drums and Uetians Triplex Model. Moreover, Pakistan Dairy Development Company has also taken an initiative to develop this kind of alternative source of energy for Pakistani farmers. Biogas is now running diesel engines, gas generators, kitchen ovens, geysers, and other utilities in Pakistan. In Nepal, the government provides subsidies to build biogas plant.

China

The Chinese had experimented the applications of biogas since 1958. Around 1970, China had installed 6,000,000 digesters in an effort to make agriculture more efficient. During the last years the technology has met high growth rates. This seems to be the earliest developments in generating biogas from agricultural waste.

In Developing Nations

Domestic biogas plants convert livestock manure and night soil into biogas and slurry, the fermented manure. This technology is feasible for small holders with livestock producing 50 kg manure per day, an equivalent of about 6 pigs or 3 cows. This manure has to be collectable to mix it with water and feed it into the plant. Toilets can be connected. Another precondition is the temperature that affects the fermentation process. With an optimum at 36 C° the technology especially applies for those living in a (sub) tropical climate. This makes the technology for small holders in developing countries often suitable.

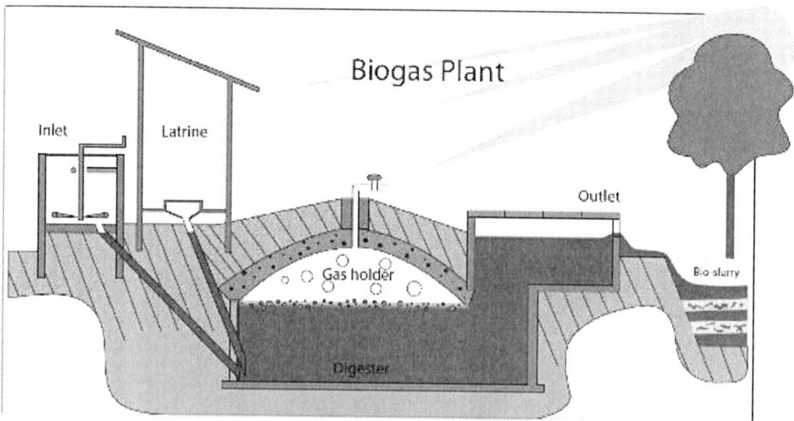

Figure: *Simple sketch of household biogas plant*

Depending on size and location, a typical brick made fixed dome biogas plant can be installed at the yard of a rural household with the investment between 300 to 500 US $ in Asian countries and up to 1400 US $ in the African context. A high quality biogas plant needs minimum maintenance costs and can produce gas for at least 15–20 years without major problems and re-investments. For the user, biogas provides clean cooking energy, reduces indoor air pollution, and reduces the time needed for traditional biomass collection, especially for women and children. The slurry is a clean organic fertilizer that potentially increases agricultural productivity.

Domestic biogas technology is a proven and established technology in many parts of the world, especially Asia. Several countries in this region have embarked on large-scale programmes on domestic biogas, such as China and India.

The Netherlands Development Organisation, SNV, supports national programmes on domestic biogas that aim to establish commercial-viable domestic biogas sectors in which local companies market, install and service biogas plants for households. In Asia, SNV is working in Nepal, Vietnam, Bangladesh, Bhutan, Cambodia, Lao PDR, Pakistan and Indonesia, and in Africa; Rwanda, Senegal, Burkina Faso, Ethiopia, Tanzania, Uganda, Kenya, Benin and Cameroon.

Solar and Wind Energy

Solar power is the conversion of sunlight into electricity, either directly using photovoltaics (PV), or indirectly using concentrated solar power (CSP). Concentrated solar power systems use lenses or mirrors and tracking systems to focus a large area of sunlight into a small beam. Photovoltaics convert light into electric current using the photoelectric effect.

Photovoltaics were initially, and still are, used to power small and medium-sized applications, from the calculator powered by a single solar cell to off-grid homes powered by a photovoltaic array. They are an important and relatively inexpensive source of electrical energy where grid power is inconvenient, unreasonably expensive to connect, or simply unavailable. However, as the cost of solar electricity is falling, solar power is also increasingly being used even in grid-connected situations as a way to feed low-carbon energy into the grid.

Commercial concentrated solar power plants were first developed in the 1980s. The 354 MW SEGS CSP installation is the largest solar power plant in the world, located in the Mojave Desert of California. Other large CSP plants include the Solnova Solar Power Station (150 MW) and the Andasol solar power station (150 MW), both in Spain.

The 250+ MW Agua Caliente Solar Project in the United States, and the 221 MW Charanka Solar Park in India, are the world's largest photovoltaic power stations.

Concentrating Solar Power

Concentrating Solar Power (CSP) systems use lenses or mirrors and tracking systems to focus a large area of sunlight into a small beam. The concentrated heat is then used as a heat source for a conventional power plant. A wide range of concentrating technologies exists: the most developed are the parabolic trough , the concentrating linear fresnel reflector, the Stirling dish and the solar power tower. Various techniques are used to track the sun and focus light. In all of these systems a working fluid is heated by the concentrated sunlight, and is then used for power generation or energy storage. Thermal storage efficiently allows up to 24 hour electricity generation.

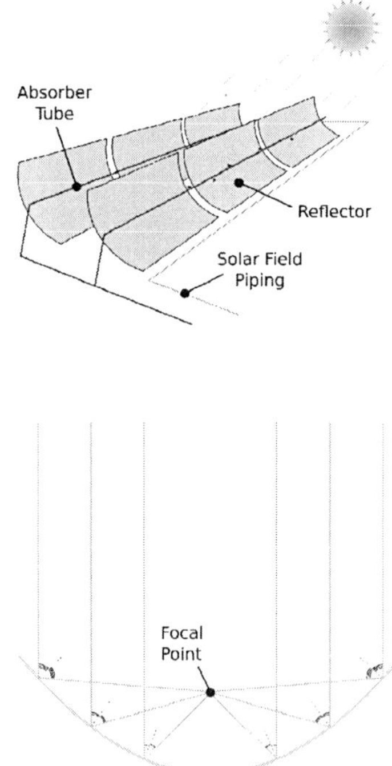

Figure: *A diagram of a parabolic trough solar farm (top), and an end view of how a parabolic collector focuses sunlight onto its focal point.*

A parabolic trough consists of a linear parabolic reflector that concentrates light onto a receiver positioned along the reflector's focal line. The receiver is a tube positioned right above the middle of the parabolic mirror and is filled with a working fluid. The reflector is made to follow the sun during daylight hours by tracking along a single axis. Parabolic trough systems provide the best land-use factor of any solar technology. The SEGS plants in California and Acciona's Nevada Solar One near Boulder City, Nevada are representatives of this technology. Compact Linear Fresnel Reflectors are CSP-plants which use many thin mirror strips instead of parabolic mirrors to concentrate sunlight onto two tubes with working fluid. This has the advantage that flat mirrors can be used which are much cheaper than parabolic mirrors, and that more reflectors can be placed in the same amount of space, allowing more of the available sunlight to be used. Concentrating linear fresnel reflectors can be used in either large or more compact plants.

The Stirling solar dish combines a parabolic concentrating dish with a Stirling engine which normally drives an electric generator. The advantages of Stirling solar over photovoltaic cells are higher efficiency of converting sunlight into electricity and longer lifetime. Parabolic dish systems give the highest efficiency among CSP technologies. The 50 kW Big Dish in Canberra, Australia is an example of this technology.

A solar power tower uses an array of tracking reflectors (heliostats) to concentrate light on a central receiver atop a tower. Power towers are more cost effective, offer higher efficiency and better energy storage capability among CSP technologies. The PS10 Solar Power Plant and PS20 solar power plant are examples of this technology.

Photovoltaics

A solar cell, or photovoltaic cell (PV), is a device that converts light into electric current using the photoelectric effect. The first solar cell was constructed by Charles Fritts in the 1880s. The German industrialist Ernst Werner von Siemens was among those who recognized the importance of this discovery. In 1931, the German engineer Bruno Lange developed a photo cell using silver selenide in place of copper oxide, although the prototype selenium cells converted less than 1% of incident light into electricity. Following the work of Russell Ohl in the 1940s, researchers Gerald Pearson, Calvin Fuller and Daryl Chapin created the silicon solar cell in 1954. These early solar cells cost 286 USD/watt and reached efficiencies of 4.5–6%.

Photovoltaic power systems

Solar cells produce direct current (DC) power which fluctuates with the sunlight's intensity. For practical use this usually requires conversion to certain desired voltages or alternating current (AC), through the use of inverters. Multiple solar cells are connected inside modules. Modules are wired together to form arrays, then tied to an inverter, which produces power at the desired voltage, and for AC, the desired frequency/phase.

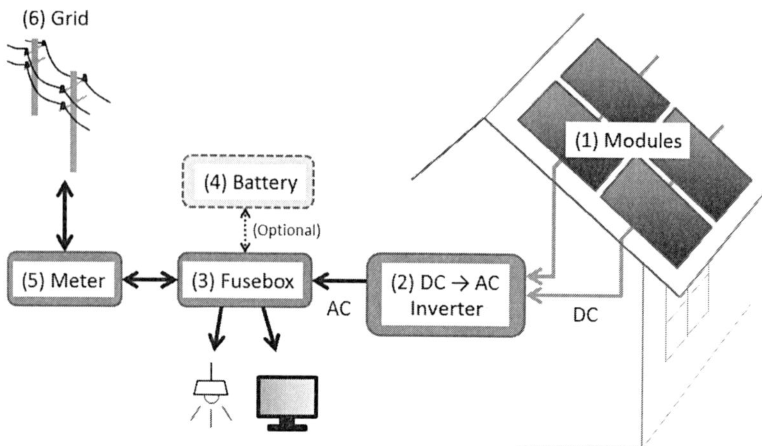

Figure: *Simplified schematics of a grid-connected residential PV power system*

Many residential systems are connected to the grid wherever available, especially in developed countries with large markets. In these grid-connected PV systems, use of energy storage is optional. In certain applications such as satellites, lighthouses, or in developing countries, batteries or additional power generators are often added as back-ups. Such stand-alone power systems permit operations at night and at other times of limited sunlight.

Development and deployment

The early development of solar technologies starting in the 1860s was driven by an expectation that coal would soon become scarce. However, development of solar technologies stagnated in the early 20th century in the face of the increasing availability, economy, and utility of coal and petroleum. In 1974 it was estimated that only six private homes in all of North America were entirely heated or cooled by functional solar power systems. The 1973 oil embargo and 1979 energy crisis caused a reorganization of energy policies around the world and brought renewed attention to developing solar technologies. Deployment strategies focused on incentive programmes such as the Federal Photovoltaic Utilization Programme in the US and the Sunshine Programme in Japan. Other efforts included the formation of research facilities in the US (SERI, now NREL), Japan (NEDO), and Germany

(Fraunhofer Institute for Solar Energy Systems ISE). Between 1970 and 1983 photovoltaic installations grew rapidly, but falling oil prices in the early 1980s moderated the growth of PV from 1984 to 1996. Since 1997, PV development has accelerated due to supply issues with oil and natural gas, global warming concerns, and the improving economic position of PV relative to other energy technologies. Photovoltaic production growth has averaged 40% per year since 2000 and installed capacity reached 39.8 GW at the end of 2010, of them 17.4 GW in Germany. As of October 2011, the largest photovoltaic (PV) power plants in the world are the Sarnia Photovoltaic Power Plant (Canada, 97 MW), Montalto di Castro Photovoltaic Power Station (Italy, 84.2 MW) and Finsterwalde Solar Park (Germany, 80.7 MW).

There are also many large plants under construction. The Desert Sunlight Solar Farm is a 550 MW solar power plant under construction in Riverside County, California, that will use thin-film solar photovoltaic modules made by First Solar. The Topaz Solar Farm is a 550 MW photovoltaic power plant, being built in San Luis Obispo County, California. The Blythe Solar Power Project is a 500 MW photovoltaic station under construction in Riverside County, California. The Agua Caliente Solar Project is a 290 megawatt photovoltaic solar generating facility being built in Yuma County, Arizona. The California Valley Solar Ranch (CVSR) is a 250 megawatt (MW) solar photovoltaic power plant, which is being built by SunPower in the Carrizo Plain, northeast of California Valley. The 230 MW Antelope Valley Solar Ranch is a First Solar photovoltaic project which is under construction in the Antelope Valley area of the Western Mojave Desert, and due to be completed in 2013.

At the end of September 2013, IKEA announced that solar panel packages for houses will be sold at 17 United Kingdom IKEA stores by the end of July 2014. The decision followed a successful pilot project at the Lakeside IKEA store, whereby one photovoltaic (PV) system was sold almost every day. The panels are manufactured by a Chinese company named Hanergy Holding Group Ltd.

Table: *Electricity Generation from Solar*

Year	*Energy (TWh)*	*% of Total*
2005	3.7	0.02%
2006	5.0	0.03%
2007	6.7	0.03%
2008	11.2	0.06%

Contd...

Year	*Energy (TWh)*	*% of Total*
2009	19.1	0.09%
2010	30.4	0.14%
2011	58.7	0.27%
2012	93.0	0.41%

Concentrating Solar Thermal Power

Commercial concentrating solar thermal power (CSP) plants were first developed in the 1980s. The 354 MW SEGS CSP installation is the largest solar power plant in the world, located in the Mojave Desert of California. Other large CSP plants include the Solnova Solar Power Station (150 MW), the Andasol solar power station (150 MW), and Extresol Solar Power Station (100 MW), all in Spain. The 370 MW Ivanpah Solar Power Facility, located in California's Mojave Desert, is the world's largest solar thermal power plant project currently under construction.

The principle advantage of CSP is the ability to efficiently add thermal storage, allowing the dispatching of electricity over up to a 24-hour period. Since peak electricity demand typically occurs at about 5 pm, many CSP power plants use 3 to 5 hours of thermal storage.

3

Seed Processing Techniques

Grains are small, hard, dry seeds (with or without attached hulls or fruit layers) harvested for human or animal food. Agronomists also call the plants producing such seeds 'grain crops'. Main types of commercial grain crops are cereals such as wheat and rye, and legumes such as beans and soybeans.

Harvested, dry grains have advantages over other staple foods such as the starchy fruits (e.g., plantains, breadfruit) and roots/tubers (e.g., sweet potatoes, cassava, yams) in the ease of storage, handling, and transport. In particular, these qualities have allowed mechanical harvest, transport by rail or ship, long-term storage in grain silos, large-scale milling or pressing, and industrial agriculture, in general. Thus, major commodity exchanges deal in canola, maize, rice, soybeans, wheat, and other grains but not in tubers, vegetables, or many other crops.

Grains—being small, hard and dry—can be stored, measured, and transported more readily than other kinds of food crops, such as fresh fruits, roots and tubers. The advent of grain agriculture allowed excess food to be produced and stored easily which could have led to the creation of the first permanent settlements and the division of society into classes.

Processing

Seed processing brings the seed to the next quality level by making high-value seed forms and seed treatments that meet the highest market standards of today.

The primary purpose of storing seeds is to save seed from one season to the next, but farmers and seed companies often find it useful or necessary to store seeds for at least two to three years, and sometimes longer.

There are several reasons for this: (1) seed yields and seed quality (germination and vigor) may be unpredictable due to growing conditions, and (2), market demand for certain crops may vary significantly from one year to the next. Market demand itself can be strongly influenced by media coverage of certain varieties that may quickly fall in or out of favour depending on media exposure. Likewise, market demand may also be influenced by the psychological effects of poor growing conditions from the previous season.

For example, gardeners coming off of two years of drought may become discouraged from making purchases for the next season. National events (such as war, disasters, or media "feeding frenzies") during peak ordering season can have a strong effect on seed sales. Cold, cloudy, winter weather, lingering late into spring can have an effect on impulse purchases by casual gardeners. Though sales to market growers tend to be more predictable from year to year, many gardeners may purchase more on impulse. Because of all these factors, seed demand is not precisely predictable. As a consequence much of the seed sold in commerce is not sold the year after it was produced. Seed is routinely carried over from year to year, and germination tested on a regular basis.

Because seed is routinely stored for more than one year, it is important to understand how seed harvesting, processing and seed storage affect the longevity and vigor of the seed. Seeds are fragile, living organisms, and the shelf life of the seed is affected at the beginning of the plant life cycle by such factors as soil nutrition. For example, if the soil is zinc deficient, the quality of the seed will be adversely affected.

(Zinc is a co-factor in many enzyme reactions in the life of the seed and the maternal plant.) Though providing the best conditions for crop growth and health is the foundation of seed quality, the factors that can have the most important effect on seed viability and vigor are harvesting, extraction, cleaning, transportation, and storage. It is easy for seed to become damaged at any of these stages. The purpose of this publication is to provide guidelines for minimizing seed damage and maximizing seed viability and vigor from pre-harvest through post-harvest processing.

Seed Harvesting and Extraction

Seed harvesting and cleaning methods can be divided into two methods: dry processing and wet processing. Dry processing involves harvesting seed that has already matured and dried within the seed-bearing portion of the plant. Examples of dry processed seed plants include beans, broccoli, corn, lettuce, okra, onions, sunflower, and turnips. Wet processing is used when the mature seed is enclosed within a fleshy fruit or berry. Examples of wet processed seed plants include cucumbers, melons, and tomatoes.

Some vegetables can be either dry processed or wet processed, for example, peppers, and squash.

Harvesting

The basic rule of harvesting is to allow the seed to mature as long as possible on the plant without the seed or fruit becoming diseased, or overly ripe. Each type of plant has an optimum time for collecting the seed, but factors such as climate, weather, disease, insects, birds, or predatory mammals may require that the seed be collected at less than the optimum time. In the Mid-Atlantic and South, frequent and daily thunderstorms and high humidity may play a large role in determining how and when seed is harvested. For example, in dry climates, beans can normally be left to mature and dry in the field, but during wet humid weather, it is best to harvest early and allow the beans to continue maturing and drying under cover.

Dry Seed Processing (Pods, Capsules, Seed Heads, Etc.)

When seeds are ready to be processed, the entire seedpod, capsule, or seed head will become brown and dry. During the maturation process, the ripening pods and capsules change colour from green, to yellow-green, to yellow, to light brown, to a darker brown, or dark gray. Ripening and maturation may be uneven within the pod or capsule, uneven on the plant, and uneven within the stand of plants. For that reason, the pods of many plants are harvested individually. Seeds of legumes and brassicas often develop a split along one side of the pod. This is the best time to collect the seed, before the pods start to open and scatter their seed. Most flower seed heads are not ready to harvest until the flower head has dried completely to the base including a short section (approximately ¼") of the supporting stem. Some plant families, such as the Asteraceae (Aster family) have a smaller percentage of viable seed in the head, and the seeds continue to mature after collection. For this reason it is best not to be too hasty in harvesting the seed. Examples include lettuce and sunflower. Some seed may mature in the capsule or pod, even before the pod has turned completely brown. Most seeds turn a darker colour as they mature.

Seeds may initially be white, turning green or tan, and then brown or black. Once the seed pods, capsules, and seed heads start to mature, it is important to check the crop on a daily basis. Rain or seed predators can ruin a good seed crop in a short period of time. Plants that produce umbels (members of the carrot family, or Umbelliferae) can usually be left in the field to harvest until the umbels are dry. Some members of this family mature their seed unevenly causing seed to scatter, while other seeds in the umbel continue to mature. One method of dealing with crops that mature their seed unevenly is to pull the plants and hang them upside down to dry under cover.

This allows the seed to continue to mature on the plant while the plant dries. This procedure is often used for lettuce. Confidence in knowing when to harvest comes both with experience and familiarity with different species and crops.

After harvest, seeds are threshed to remove the seed from the surrounding plant material. A period of air-drying is important before seeds are threshed. Plant material should be spread out in thin layers until all plant material is dry; otherwise, mold, decay, and heat from decay will cause damage to the seeds. As the plant material dries, seed pods may split open or shed seed. Harvested material should be stored in a well-ventilated room with low humidity. During this time you should be aware of insects, especially weevils that feed on the seeds. Plant material that is ready to be threshed should be brittle. Threshing is best done outside on a dry day. The threshing process involves application of mechanical force using a controlled pressure and a shearing motion., and is accomplished by hand or by machine.

There are many different methods for threshing seed. Plants that have pods, such as beans and okra, can be threshed by placing the pods in a large feed sack, which is tied shut securely, and then placed on the ground where it is flailed, stepped on, jogged on, or danced on with a twisting motion. The sack is turned often to redistribute the plant material for further threshing. When using this method, it is best to use running shoes or other soft-soled shoes because seeds can develop hairline cracks and splits from too much pressure. To separate seed from flower heads, plant material is spread on a concrete slab and then gently walked on with a twisting motion to break open the flower heads. Care must be taken to not apply so much pressure that the seed is abraded or broken (especially a concern with angular-shaped seed). To thresh seeds of brassicas, place plant material in a large wheelbarrow or on a large tarp, and while wearing gloves twist and wring the plant material through your hands until the seed breaks free from the pods. Another method for extracting seed is to place seed heads on a piece of plywood. The seed is extracted by placing a wooden cement float above the material and then pressing and twisting until the seed breaks free. If the seed heads are small you can run a rolling pin over them. For threshing small lots of seed, a threshing box may be used. This consists of a wooden box, with sides slanting outward, open at the top and on one end, and the long sides and bottom covered with corrugated-rubber floor matting. Seed is placed in the box and a rubbing board (also covered with corrugated rubber), is moved back and forth across the seed heads to separate the seed from the heads.

Wet seed processing (crops with fleshy fruits, fermentation): Wet seed processing is used with seed crops that have seeds in fleshy fruits or berries. There are three steps to the process: (1) extraction of the seed from the fruit, (2) washing the seeds, and (3), drying.

Extracting Seed

The type of extraction process depends on the species. Soft fruits such as tomatoes are cut up, mashed, and then fermented. Cucumbers and melons are cut in half, the seed scraped out along with the fruit pulp surrounding the seed, and then fermented. In watermelons, the entire fleshy fruit is fermented along with the extracted seed. These types of fruits have a gel surrounding the seed that contains germination inhibitors. The presence of the gel also makes handling and drying of the seed difficult. Fermentation is a natural process that occurs to a small extent as fruits decompose.

When fermentation is done in a controlled manner, the microorganisms, principally yeast, break down the gel thus releasing the seed while killing bacteria and fungi that cause most seed-borne diseases. The temperature and length of fermentation are important. If the mash is not fermented long enough, seed-borne diseases will not be eliminated, but if fermented too long, the seeds may sprout prematurely. The length of the fermentation is dependent on temperature and typically last three days at a temperature of 70 to 75oF (21oC to 24oC). Length of fermentation may also depend on the variety itself. For example, varieties with high sugar content may take longer to ferment, up to four days. With few exceptions, fermentation periods longer than three days risk damaging the seed. There are different fermentation techniques for different crops, for example, pepper seeds are extracted from the fruits by mashing, but the fermentation process may last only 24 to 48 hours. Though eggplant isn't a watery fruit, it can be mashed and fermented for about 48 hours. The grinding plates on the mill are set wide enough apart to break up the fruit without grinding the seed. Small peppers can be put directly into the mill, whereas larger fruits such as eggplant are cut into small cubes before processing. Before processing another variety, the mill has to be disassembled and cleaned thoroughly to remove bits of small seed caught in the plates and other parts of the mill. Though some people use a food processor and blender set on low speed to extract the seed of hot peppers and eggplant, it is easy to damage the seed. I would not recommend using a blender. Instead, a food processor that has a thick (approximately 1/8") blade works well provided there is ample water in the processor, the speed is as slow as possible, and the processing time as short as possible.

There are some differences of opinion about adding water to the fermenting mixture, the concern being that water slows the fermentation process causing premature sprouting. In my experience this hasn't been a problem. As long as the ferment is not diluted more than perhaps 10 to 20 percent by volume, this isn't a concern. The issue of whether to add water depends on how thick the mash is and it depends on the variety. Some varieties make a very thick mash that is hard to stir, and others make a watery mash that stirs easily. The mash should be stirred three times a day, once in the morning, once in mid-day, and once in the evening. If the mash is too thick to stir easily, the nutrients are not going to circulate easily. Stirring is also important for better control of seed-borne diseases. In addition, when the mash is not stirred, a foul-smelling, white mold forms on top of the mash. This mold can discolour (darken) and damage seeds at the top of the mash. These seeds will later have to be hand picked out of the dried seed. A properly fermenting mash should not have a foul smell, and there should be little or no white mold on the top. A small amount of white mold is not harmful and can be stirred back into the mash, but a heavy overgrowth should be removed. !

Washing Seed

After fermentation is complete, the seeds are washed to remove pulp, pieces of fruit and debris, and low quality seed. Before washing the seed, it is useful (especially for washing tomato seed) to first scoop out pieces of pulp floating on top of the mash. This is done by straining the mash with your fingers, pulling out the larger chunks. Whether or not there is floating pulp depends on the variety or how thoroughly the fruit was processed. Add a volume of water at least equal twice the volume of mash. It is important to dilute the mash sufficiently because the more dissolved solids there are in the mash, the higher the specific gravity. If the specific gravity is high (lots of soluble solids) it will be more difficult to wash the seeds properly. As a general rule, good seeds are heavy and sink to the bottom, whereas poor quality seeds are light and tend to float off with the wash. The washing process is repeated until the wash water becomes clear. Although most good seeds sink to the bottom, some vegetables have very light seed and require extra care during the washing process. For example, it is common for a significant amount of good pepper seed to float rather than sink during the washing. This can be avoided by adding the wash water slowly, so as not to create tiny air bubbles that adhere to the seeds, making them buoyant. Even with this precaution, there are a few varieties of watermelon for example, where the good seeds tend to float rather than sink, thereby requiring special care in washing. In this case, the best way to wash the seed is to pour the wash through a ¼" hardware cloth screen, and then use a hose to force pulp through the screen.

Drying Seed

Seeds should be dried fairly quickly after washing. Slow drying may result in mold growth or premature sprouting of the seed. In the Mid-Atlantic and South, seeds should not be dried in the sun, nor should they be dried anywhere where the temperature exceeds 95oF (35oC). Dark coloured seeds are especially vulnerable to damage when sun dried. Instead, seeds should be dried in a climate-controlled environment using fan ventilation. A combination of ceiling fans and air conditioning dries seed safely and very quickly. Seed should be spread out in thin layers (no thicker than ¼" for small seeds) and then stirred several times a day until dry. Once the seeds feel dry, they should cure for another two to three weeks. Curing is the final stage in the drying process. As the seed moisture content declines it comes into equilibrium with the relative humidity. After the seeds are cured they can be placed in a container.

When drying seeds, choose plywood, window screen, or any hard, non-stick surface. Avoid using paper towels, newspaper, cardboard, or cloth because seeds will tend to stick to the surface making them difficult to remove. Beginning growers often make the mistake of drying squash seed on newspaper, which adheres permanently to the seed coat. No one wants to read the daily news on the surface of squash seed.

Seed Processing and Handling

Threshing: Equipment

For small-scale seed production, it is often not cost effective to purchase threshing equipment because of the small volume of seed produced. A number of small alternative seed companies, specializing in heirlooms and speciality seed, thresh most, if not all of their seed by hand. Some seed crops, especially herb and flowers seeds are produced in such small quantity (and have such small seeds) that it is impractical to use a machine for threshing. Also flowers and other seed crops, if left to dry on canvas or a tarp, naturally dehisce (shed) much of their seed. The remaining amount of seed can be released by flailing or rolling, or other non-mechanical means.

It is difficult to locate low-cost, low-tech shredders for small-scale seed threshing. The options are to locate old, used equipment or to construct your own. Vegetable seeds, such as beans can be threshed with a modified leaf shredder/chipper. Typical threshing rates for this machine are 22 to 30 pounds/hour for brassicas, and approximately 100 pounds/hour for beans. The research and development group at the Organic Gardening Experimental Farm of Rodale Press developed a small-scale thresher designed to thresh a variety of crops ranging from amaranth to soybeans. Though the machine was not fully refined it did an acceptable job.

Seed Cleaning Equipment (Seed Screens)

It is not necessary to have expensive seed cleaning equipment to clean seed for small-scale production.

With the exception of lettuce seed and a few others , the majority of crop seeds can be cleaned with homemade seed screens. Winnowing will still be necessary to remove smaller chaff.

Many seeds can be screened with several different mesh sizes of hardware cloth. Hardware cloth is readily available in the following mesh sizes: 1/2", 3/8", 1/4", and 1/8". The 1/2" and 1/4" mesh sizes are available at most hardware stores. The 3/8" and 1/8" mesh sizes are the most useful sizes, but are the most difficult to find.

Aside from hardware cloth, a lot of other materials are useful for making seed screens. Aluminum window screen can be used for small seeds. You'll also find various meshes of screen available in the housewares section of department stores. Cabinets from electronic devices often have round or oblong holes which are useful for cleaning some types of seed. Special meshes can be ordered from mail-order hardware speciality catalogue. It is useful to collect a large variety of mesh sizes and shapes to handle a wide variety of seed types.

Once you have located suitable screen material it should be mounted on a frame. Professional seed screens are mounted on 12" square wooden frames. Homemade screens should also be mounted on 12" square frames so they can be used together with professional screens. The frames can be made of wooden lath which measures 3/4" x 1-1/2". Ideally, the screens should be constructed so that they nest together.

The nesting feature is desirable for using three screens simultaneously: rough chaff is retained on the top screen, seeds in the middle screen, and small chaff and small seeds on the lower screen. Once the frame has been made, the screen is nailed with small nails or brads at approximately 1" intervals all along the edge of the frame.

Winnowing Equipment

The classic method of winnowing involves placing seeds in a wide basket and tossing the seeds and chaff into the air. The chaff is carried away by the wind. Illustrations of Native Americans using this historic method evoke a bucolic mood, but in reality, winnowing by this method is extremely difficult, and the results are not very satisfactory. The most vexing part of the process is that wind speed is always changing in velocity and direction. It can work for certain kinds of seed, but it actually works better to use two large bowls, pouring seed from one bowl into another bowl below, while blowing on the plant material as it falls.

This method works satisfactorily if the seed is heavy and the chaff is very fine, and susceptible to being carried away by a gentle current of air. In any case, hand winnowing should be done, not on windy days, but when the air is calm.

A good range of equipment for winnowing includes the following: (1) an assortment of stainless steel bowls ranging in size from 6 to 16" in diameter, preferably bowls with varying shallow and deep sides; (2) an electric hair dryer with two speeds (heating element removed, or unheated air setting); (3) portable vacuum cleaner with option of connecting hose to the air discharge opening: (4) household box fan with three speeds; (5) squirrel cage blower; (6) rheostat for controlling fan motor speed; (7) a tarp for catching seed or chaff; and (8) a dust mask for keeping chaff out of your lungs.

Most light seed can be winnowed quickly and efficiently with mixing bowls and a hair dryer. It takes a little practice determining the correct amount of seed to put in the bowl, setting the proper fan speed, and the distance of the dryer from the bowl. It helps to jiggle the bowl up and down or to swirl the seed in the bowl. There must be little or no wind when you work with light seed. Until you have a little practice it is best to put a tarp on the ground, in case you need to sweep up your mistake and start over. Heavy seed is best winnowed by pouring the seed from one container to another in front of box fan, or alternatively directing the air discharge hose from a vacuum cleaner into a large (16" diameter) mixing bowl.

When dealing with large volumes or certain types of seed it is helpful to use mechanical equipment for winnowing.

Seed Drying

Principles of Drying

Drying is a normal part of the seed maturation process. Some seeds must dry down to minimum moisture content before they can germinate. Low seed moisture content is a pre-requisite for long-term storage, and is the most important factor affecting longevity. Seeds lose viability and vigor during processing and storage mainly because of high seed moisture content (seed moisture greater than 18%).

High seed moisture causes a number of problems:

- Moisture increases the respiration rate of seeds, which in turn raises seed temperature. For example, in large-scale commercial seed storage, respiring seeds may generate enough heat to kill the seeds quickly, or to even start a fire if not dried sufficiently. Small-scale growers are not likely to have such an extreme condition, but seed longevity will, nevertheless be affected.
- Mold growth will be encouraged by moisture, damaging the seeds either slowly or quickly, depending on the moisture content of the seeds. Some molds that don't grow well at room temperature may grow well at low temperatures causing damage to refrigerated seeds. In such a case there may be no visual sign of damage.
- Unless seed moisture is at least eight percent or below, insects such as weevils can breed causing rapid destruction of seeds in a short period of time.

Drying Seeds for Long-term Storage

Silica gel is the most effective desiccant (moisture absorbing material) for drying seeds. Powdered milk has been recommended as a desiccant in older seed-saving literature, but is less than ten percent as effective as silica gel. Silica gel is a highly porous form of silica that absorbs moisture. It is available as a powder or as beads in different sizes. The best size bead for drying seeds is approximately 1/16" to 1/8" in diameter "Colour-indicating" silica gel is a form of silica gel that has been treated with a small amount of cobalt chloride which acts as a moisture indicator. When the indicator gel is completely dry it is a deep blue colour.

As the gel absorbs moisture from the air, it gradually changes in colour from deep blue to light pink.

Though silica gel is clearly allowable under the Organic Rules, colour-indicating silica gel is not clearly covered. Check with your certifier before using colour-indicating silica gel.

Silica gel can be repeatedly reactivated (re-dried) after it has absorbed moisture. The procedure involves heating the gel and driving off the moisture, and as it dries the colour gradually changes to deep blue. Drying must be done in a controlled fashion, otherwise the beads will turn black, and the moisture-absorbing capacity of the beads will be destroyed. There are two methods for reactivating silica gel:

Oven-drying method

This method gives the best results, but it takes longer and uses more energy. Set the oven for a temperature of at least 200oF, but no higher than 275oF. Remember, that some ovens may run hotter than the set temperature. Place the silica gel in a thick-walled Pyrex dish, no deeper than one inch, and continue heating until the beads turn deep blue. When drying large quantities (a pound or more), the gel should be stirred occasionally. The oven drying method takes 1-1/2 hours per quart of gel (with the oven temperature set at 275oF). One quart of silica gel weighs approximately 30 ounces (1.9 pounds).

Microwave-drying Method

The microwave method works much faster but must be monitored more closely to avoid overheating the gel. Use only a thick-walled Pyrex container for heating the gel. Set the microwave on medium or medium high and dry for approximately three to five minutes. The colour change in the gel can be monitored through the over door, but the gel should be inspected and stirred at the end of each heating cycle. If the gel has not dried, heat again for another three to five minutes. Approximate drying time is eight to twelve minutes per pound of gel, though actual heating time will depend according to the type of microwave.

There are some fine points and cautionary notes about using silica gel. When using a Pyrex dish, the glass should be thick. Do not use plastic microwave containers that will melt on contact with the gel.

Silica gel gets very hot, and the glass container may shatter if it is too thin, of the wrong type, or if unevenly heated. You may notice a slight odor during heating. This is due to either overheating the silica gel, or to the evaporation of organic seed volatiles absorbed by the silica gel during the drying process. Silica gel itself is chemically inert, non-toxic, non-corrosive, and odorless, but breathing the dust can be hazardous under prolonged and repeated exposure. For that reason, do not use the finely ground or powdered form.

Instructions for drying seed with silica gel:

- To dry seed, determine the weight of the seed to be dried, including the packets or envelopes that contain the seed. Measure out an equal weight of silica gel and place the seeds and silica gel in an airtight container for seven days. When drying seed it is important to keep the container size small in relation to the volume of seeds being dried.
- At the end of seven days, remove the packets of seed from the drying container and transfer into another airtight container, such as a Mason jar, Seed Saver Vial™ or barrier pouch. Because seeds can re-absorb moisture from the air quickly, they should be transferred quickly.

When used as directed, silica gel dries seed from 12% typical moisture content to a desired moisture content of approximately 5% for small seeds and 7% for large seeds. As a "rule of thumb", seeds will not be damaged provided the drying time doesn't exceed seven days. This drying time applies to humid climates, such as the Mid-Atlantic. Longer drying may drop the moisture content below 3% for small seeds and below 5% for large seeds, levels which may damage seeds or force them into dormancy. Legumes, such as beans, are especially are injured by over drying.

Seed Storage

Effect of Temperature on Seed Longevity: The general effect of temperature on longevity is that longevity increases as temperature decreases. This is true of "orthodox" seeds: that is, most seeds that follow some general "rules of thumb" regarding longevity during the storage life of the seeds. The relationship between temperature and seed longevity is that for each 10oF (5.6oC) decrease in temperature, longevity doubles (Harrington, 1972). This rule applies to seeds stored between temperatures of 32oF (0oC) and 122oF (50oC). This rule assumes that the moisture content is a constant. This is a general guideline; in reality the longevity of some vegetable species declines more rapidly than suggested by the rule, while the longevity of others declines more slowly in relation to storage temperature.

The longevity of seeds is generally not affected by subfreezing temperatures provided the moisture content is less than 14% (because ice crystals do not form). This has been established by a number of published reports as well as my own experience storing and germination testing a wide variety of vegetable, flower, and herb seeds at 20oF (-7oC) and below for periods of at least five years. I am still getting excellent germination from seed stored twenty years, which was dried to approximately five percent moisture content.

This is the ideal way to store seed, especially small seed that doesn't require much freezer space. One caveat: seed cycled in and out of the freezer too many times without redrying may cause degradation of germination.

Effect of Seed Moisture and Humidity on Seed Longevity: Seed moisture has a greater effect than temperature on seed longevity (as noted in the previous section titled "principles of seed drying"). Most seeds also follow some "rules of thumb" regarding moisture and longevity. The general relationship is that for each one percent increase in seed moisture, longevity decreases by half (Harrington, 1972). This rule applies to seed with moisture content between 5 and 13%.

Above 13% moisture content, seed storage fungi and increased heating due to respiration cause longevity to decline at a faster rate. Once seed moisture reaches 18 to 20%, increased respiration, and the activity of microorganisms cause rapid deterioration of the seed. At 30% moisture content, most non-dormant seeds germinate. At the low end of the moisture range, seed stored at 4 to 5% moisture content is unaffected by seed storage fungi, but such seeds have a shorter longevity than seed stored at a slightly higher moisture content (Bewley and Black, 1985).

Relationship Between Relative Humidity and Seed Moisture Content: When storing commercially grown seed, it is impractical and too costly to use desiccant to dry the seed for storage, unless the seed is small and expensive. Commercial seed is usually packaged for short or long-term storage under conditions of ambient humidity (unless special equipment is used). Because relative humidity has a significant effect on seed moisture content, it is important to understand the relationship between humidity and seed moisture.

Regardless of the type of storage conditions, the moisture content of seed eventually comes into equilibrium with the moisture in the surrounding air. The relationship between atmospheric relative humidity and seed moisture content.

Seed Treatment

In agriculture and horticulture, a seed treatment or seed dressing is a chemical, typically antimicrobial or fungidal, with which seeds are treated (or "dressed") prior to planting. Less frequently insecticides are added. Seed treatments can be an environmentally more friendly way of using pesticides as the amounts used can be very small. It is usual to add colour to make treated seed less attractive to birds if spilt and easier to see and clean up in the case of an accidental spillage.

One seed treatment, imidacloprid, from the neonicotinoid family of insecticides, is controversial and was banned in France for use on maize, due to that government's belief that the chemical was implicated in recent dramatic drops in bee counts, and possibly in Colony Collapse Disorder. Dust from treated seed is known to have caused at least some problems particularly from crops such as maize drilled during the main honey flows. Improvements to pneumatic drills to reduce dust release, and improvements to seed treatment compounds to prevent the compound breaking up into dust have been introduced in Europe led by Germany and Holland from 2009 to 2012. Seed coating is a thicker form of covering of seed and may contain fertilizer, growth promoters and or seed treatment as well as an inert carrier and a polymer outer shell.

Seed dressing is also used to refer to the process of removing chaff, weed seeds and straw from a seed stock. Care is needed not to confuse the two.

Other Methods of Processing

Freezing: Most vegetables should be blanched before freezing to prevent loss of flavour and colour during storage. Freezing temperatures are best set between -21 to -18 C (0 to 5 F).

Packages for freezing should be moisture proof and vapor proof and contain as little air as possible to prevent oxidation during storage. Heavy plastic bags, heavy aluminum foil, glass freezer jars and waxed freezer cartons all make good containers.

Jellies, Jams and Preserves

Making jams, jellies and other high sugar preserves requires a balance of fruit, acid, pectin and sugar for best results. Underripe fruits contain more pectin than ripe fruits, and apple juice is a good source of natural pectin. If fruits are low in acid, lemon juice can be added to the mixture of fruit and sugar. Cane or beet sugar is better for making preserves than honey or corn syrup.

To preserve fruits, cook on medium heat until the mixture "sheets" from a spoon. Avoid overcooking since this will lower the jelling capacity of the mixture. Pour into containers and seal with paraffin wax (jellies only). The other preserves should be processed in a boiling water bath for five minutes.

Fermentation

When lactic acid bacteria in foods convert carbohydrates to lactic acid, food is preserved by the resulting low pH. Sauerkraut (cabbage) and wine (grapes) are two examples of thousands of fermented foods made around the world.

Acidification

Pickling is a simple processing method that can be used with many types of fruits and vegetables. Brine solution (9 parts cider or white vinegar, 1 part non-iodized salt, 9 parts water, plus flavourings and spices) is poured over the product into glass canning jars (leave 1/2 inch headspace). Brined pickles are sealed and left at ambient temperature for three or more weeks, while fresh pack pickles are processed in a boiling water bath for 10 minutes.

4

Fruit Specific Processing Technologies

Fruits are the means by which these plants disseminate seeds. Many of them that bear edible fruits, in particular, have propagated with the movements of humans and animals in a symbiotic relationship as a means for seed dispersal and nutrition, respectively; in fact, humans and many animals have become dependent on fruits as a source of food. Fruits account for a substantial fraction of the world's agricultural output, and some (such as the apple and the pomegranate) have acquired extensive cultural and symbolic meanings.

In common language usage, "fruit" normally means the fleshy seed-associated structures of a plant that are sweet or sour and edible in the raw state, such as apples, oranges, grapes, strawberries, bananas, and lemons. On the other hand, the botanical sense of "fruit" includes many structures that are not commonly called "fruits", such as bean pods, corn kernels, wheat grains, and tomatoes.

Figure: *A fruit results from maturation of one or more flowers, and the gynoecium of the flower(s) forms all or part of the fruit.*

Inside the ovary/ovaries are one or more ovules where the megagametophyte contains the egg cell. After double fertilization, these ovules will become seeds. The ovules are fertilized in a process that starts with pollination, which involves the movement of pollen from the stamens to the stigma of flowers. After pollination, a tube grows from the pollen through the stigma into the ovary to the ovule and two sperm are transferred from the pollen to the megagametophyte. Within the megagametophyte one of the two sperm unites with the egg, forming a zygote, and the second sperm enters the central cell forming the endosperm mother cell, which completes the double fertilization process. Later the zygote will give rise to the embryo of the seed, and the endosperm mother cell will give rise to endosperm, a nutritive tissue used by the embryo.

As the ovules develop into seeds, the ovary begins to ripen and the ovary wall, the *pericarp*, may become fleshy (as in berries or drupes), or form a hard outer covering (as in nuts). In some multiseeded fruits, the extent to which the flesh develops is proportional to the number of fertilized ovules. The pericarp is often differentiated into two or three distinct layers called the *exocarp* (outer layer, also called epicarp), *mesocarp* (middle layer), and *endocarp* (inner layer). In some fruits, especially simple fruits derived from an inferior ovary, other parts of the flower (such as the floral tube, including the petals, sepals, and stamens), fuse with the ovary and ripen with it. In other cases, the sepals, petals and/or stamens and style of the flower fall off. When such other floral parts are a significant part of the fruit, it is called an *accessory fruit.* Since other parts of the flower may contribute to the structure of the fruit, it is important to study flower structure to understand how a particular fruit forms.

There are three general modes of fruit development:

- Apocarpous fruits develop from a single flower having one or more separate carpels, and they are the simplest fruits.
- Syncarpous fruits develop from a single gynoecium having two or more carpels fused together.
- Multiple fruits form from many different flowers.

Plant scientists have grouped fruits into three main groups, simple fruits, aggregate fruits, and composite or multiple fruits. The groupings are not evolutionarily relevant, since many diverse plant taxa may be in the same group, but reflect how the flower organs are arranged and how the fruits develop.

Simple Fruit

Figure: *Epigynous berries are simple fleshy fruit. Clockwise from top right: cranberries, lingonberries, blueberries red huckleberries*

Simple fruits can be either dry or fleshy, and result from the ripening of a simple or compound ovary in a flower with only one pistil. Dry fruits may be either dehiscent (opening to discharge seeds), or indehiscent (not opening to discharge seeds). Types of dry, simple fruits, with examples of each, are:

- achene - Most commonly seen in aggregate fruits (e.g. strawberry)
- capsule – (Brazil nut)
- caryopsis – (wheat)
- Cypsela - An achene-like fruit derived from the individual florets in a capitulum (e.g. dandelion).
- fibrous drupe – (coconut, walnut)
- follicle – is formed from a single carpel, and opens by one suture (e.g. milkweed). More commonly seen in aggregate fruits (e.g. magnolia)
- legume – (pea, bean, peanut)
- loment - a type of indehiscent legume
- nut – (hazelnut, beech, oak acorn)
- samara – (elm, ash, maple key)
- schizocarp – (carrot seed)
- silique – (radish seed)

- silicle – (shepherd's purse)
- utricle – (beet)

Fruits in which part or all of the *pericarp* (fruit wall) is fleshy at maturity are *simple fleshy fruits*. Types of fleshy, simple fruits (with examples) are:

- berry – (redcurrant, gooseberry, tomato, cranberry)
- stone fruit or drupe (plum, cherry, peach, apricot, olive)

An aggregate fruit, or *etaerio*, develops from a single flower with numerous simple pistils.

- Magnolia and Peony, collection of follicles developing from one flower.
- Sweet gum, collection of capsules.
- Sycamore, collection of achenes.
- Teasel, collection of cypsellas
- Tuliptree, collection of samaras.

The pome fruits of the family Rosaceae, (including apples, pears, rosehips, and saskatoon berry) are a syncarpous fleshy fruit, a simple fruit, developing from a half-inferior ovary.

Schizocarp fruits form from a syncarpous ovary and do not really dehisce, but split into segments with one or more seeds; they include a number of different forms from a wide range of families. Carrot seed is an example.

Aggregate Fruit

Aggregate fruits form from single flowers that have multiple carpels which are not joined together, i.e. each pistil contains one carpel. Each pistil forms a fruitlet, and collectively the fruitlets are called an etaerio. Four types of aggregate fruits include etaerios of achenes, follicles, drupelets, and berries. Ranunculaceae species, including *Clematis* and *Ranunculus* have an etaerio of achenes, *Calotropis* has an etaerio of follicles, and *Rubus* species like raspberry, have an etaerio of drupelets. *Annona* have Etaerio of berries.

The raspberry, whose pistils are termed *drupelets* because each is like a small drupe attached to the receptacle. In some bramble fruits (such as blackberry) the receptacle is elongated and part of the ripe fruit, making the blackberry an *aggregate-accessory* fruit. The strawberry is also an aggregate-accessory fruit, only one in which the seeds are contained in achenes. In all these examples, the fruit develops from a single flower with numerous pistils.

Multiple Fruits

A multiple fruit is one formed from a cluster of flowers (called an *inflorescence*). Each flower produces a fruit, but these mature into a single mass. Examples are the pineapple, fig, mulberry, osage-orange, and breadfruit.

In the photograph on the right, stages of flowering and fruit development in the noni or Indian mulberry (*Morinda citrifolia*) can be observed on a single branch. First an inflorescence of white flowers called a head is produced. After fertilization, each flower develops into a drupe, and as the drupes expand, they become *connate* (merge) into a *multiple fleshy fruit* called a *syncarpet.*

Berries

Berries are another type of fleshy fruit; they are simple fruit created from a single ovary. The ovary may be compound, with several carpels. Type include (examples follow in the table below):

- Pepo – Berries where the skin is hardened, cucurbits
- Hesperidium – Berries with a rind and a juicy interior, like most citrus fruit

Accessory Fruit

Some or all of the edible part of accessory fruit is not generated by the ovary.

Table of Fruit Examples

Types of fleshy fruits					
True berry	*Pepo*	*Hesperidium*	*Aggregate fruit*	*Multiple fruit*	*Accessory fruit*
Blackcurrant, Redcurrant, Gooseberry, Tomato, Eggplant, Guava, Lucuma, Chili pepper, Pomegranate, Kiwifruit, Grape, Cranberry, Blueberry	Pumpkin, Gourd, Cucumber, Melon	Orange, Lemon, Lime, Grapefruit	Blackberry, Raspberry, Boysenberry	Pineapple, Fig, Mulberry, Hedge apple	Apple, Rose hip, Strawberry

Seedless Fruits

Seedlessness is an important feature of some fruits of commerce. Commercial cultivars of bananas and pineapples are examples of seedless fruits. Some cultivars of citrus fruits (especially navel oranges), satsumas, mandarin oranges, table grapes, grapefruit, and watermelons are valued for their seedlessness. In some species, seedlessness is the result of *parthenocarpy*, where fruits set without fertilization. Parthenocarpic fruit set may or may not require pollination but most seedless citrus fruits require stimulus from pollination to produce fruit.

Seedless bananas and grapes are triploids, and seedlessness results from the abortion of the embryonic plant that is produced by fertilization, a phenomenon known as *stenospermocarpy* which requires normal pollination and fertilization.

Seed Dissemination

Variations in fruit structures largely depend on the mode of dispersal of the seeds they contain. This dispersal can be achieved by animals, wind, water, or explosive dehiscence.

Some fruits have coats covered with spikes or hooked burrs, either to prevent themselves from being eaten by animals or to stick to the hairs, feathers or legs of animals, using them as dispersal agents. Examples include cocklebur and unicorn plant.

The sweet flesh of many fruits is "deliberately" appealing to animals, so that the seeds held within are eaten and "unwittingly" carried away and deposited at a distance from the parent. Likewise, the nutritious, oily kernels of nuts are appealing to rodents (such as squirrels) who hoard them in the soil in order to avoid starving during the winter, thus giving those seeds that remain uneaten the chance to germinate and grow into a new plant away from their parent.

Other fruits are elongated and flattened out naturally and so become thin, like wings or helicopter blades, e.g. maple, tuliptree and elm. This is an evolutionary mechanism to increase dispersal distance away from the parent via wind. Other wind-dispersed fruit have tiny *parachutes*, e.g. dandelion and salsify.

Coconut fruits can float thousands of miles in the ocean to spread seeds. Some other fruits that can disperse via water are nipa palm and screw pine.

Some fruits fling seeds substantial distances (up to 100 m in sandbox tree) via explosive dehiscence or other mechanisms, e.g. impatiens and squirting cucumber.

Uses

Many hundreds of fruits, including fleshy fruits like apple, peach, pear, kiwifruit, watermelon and mango are commercially valuable as human food, eaten both fresh and as jams, marmalade and other preserves. Fruits are also used in manufactured foods like cookies, muffins, yogurt, ice cream, cakes, and many more. Many fruits are used to make beverages, such as fruit juices (orange juice, apple juice, grape juice, etc.) or alcoholic beverages, such as wine or brandy. Apples are often used to make vinegar. Fruits are also used for gift giving, Fruit Basket and Fruit Bouquet are some common forms of fruit gifts.

Many vegetables are botanical fruits, including tomato, bell pepper, eggplant, okra, squash, pumpkin, green bean, cucumber and zucchini. Olive fruit is pressed for olive oil. Spices like vanilla, paprika, allspice and black pepper are derived from berries.

Nutritional Value

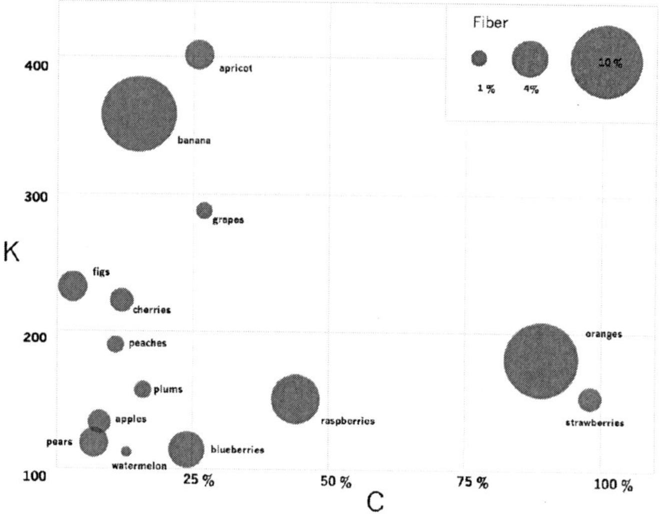

Each point refers to a 100 g serving of the fresh fruit, the daily recommended allowance of vitamin C is on the X axis and mg of Potassium (K) on the Y (offset by 100 mg which every fruit has) and the size of the disk represents amount of fibre (key in upper right). Oranges and bananas have the highest amount of fibre and are also furthest away from the bottom left indicating their superior nutrition. Watermelon which has almost no fibre and low levels of vitamin C and potassium comes in last place.

Fruits are generally high in fibre, water, vitamin C and sugars, although this latter varies widely from traces as in lime, to 61% of the fresh weight of the date. Fruits also contain various phytochemicals that do not yet have an RDA/RDI listing under most nutritional factsheets, and which research indicates are required for proper long-term cellular health and disease prevention. Regular consumption of fruit is associated with reduced risks of cancer, cardiovascular disease (especially coronary heart disease), stroke, Alzheimer disease, cataracts, and some of the functional declines associated with aging.

Diets that include a sufficient amount of potassium from fruits and vegetables also help reduce the chance of developing kidney stones and may help reduce the effects of bone-loss. Fruits are also low in calories which would help lower one's calorie intake as part of a weight-loss diet.

Nonfood Uses

Because fruits have been such a major part of the human diet, different cultures have developed many different uses for various fruits that they do not depend on as being edible. Many dry fruits are used as decorations or in dried flower arrangements, such as unicorn plant, lotus, wheat, annual honesty and milkweed. Ornamental trees and shrubs are often cultivated for their colourful fruits, including holly, pyracantha, viburnum, skimmia, beautyberry and cotoneaster.

Fruits of opium poppy are the source of opium which contains the drugs morphine and codeine, as well as the biologically inactive chemical theabaine from which the drug oxycodone is synthesized. Osage orange fruits are used to repel cockroaches. Bayberry fruits provide a wax often used to make candles. Many fruits provide natural dyes, e.g. walnut, sumac, cherry and mulberry. Dried gourds are used as decorations, water jugs, bird houses, musical instruments, cups and dishes. Pumpkins are carved into Jack-o'-lanterns for Halloween. The spiny fruit of burdock or cocklebur were the inspiration for the invention of Velcro.

Coir is a fibre from the fruit of coconut that is used for doormats, brushes, mattresses, floortiles, sacking, insulation and as a growing medium for container plants. The shell of the coconut fruit is used to make souvenir heads, cups, bowls, musical instruments and bird houses.

Fruit is often used as a subject of still life paintings.

Safety

For food safety, the CDC recommends proper fruit handling and preparation to reduce the risk of food contamination and foodborne illness. Fresh fruits and vegetables should be carefully selected. At the store, they should not be damaged or bruised and pre-cut pieces should be refrigerated or surrounded by ice. All fruits and vegetables should be rinsed before eating. This recommendation also applies to produce with rinds or skins that are not eaten. It should be done just before preparing or eating to avoid premature spoilage. Fruits and vegetables should be kept separate from raw foods like meat, poultry, and seafood, as well as utensils that have come in contact with raw foods. Fruits and vegetables, if they are not going to be cooked, should be thrown away if they have touched raw meat, poultry, seafood or eggs. All cut, peeled, or cooked fruits and vegetables should be refrigerated within two hours. After a certain time, harmful bacteria may grow on them and increase the risk of foodborne illness.

Production

Most fruit is produced using traditional farming practices. However, the yield of fruit from organic farming is growing.

Storage

The plant hormone ethylene causes ripening of many types of fruit. Maintaining most fruits in an efficient cold chain is optimal for post harvest storage, with the aim of extending and ensuring shelf life. All fruits benefit from proper post harvest care.

Processing

Fruit Quality

Fruit quality goes back to tree stock, growing practices and weather conditions. Closer to the shipper and processor, however, are the degrees of maturity and ripeness when picked and the method of picking or harvesting.

There is a distinction between maturity and ripeness of a fruit. Maturity is the condition when the fruit is ready to eat or if picked will become ready to eat after further ripening. Ripeness is that optimum condition when colour, flavour and texture have developed to their peak.

Some fruit is picked when it are mature but not yet ripe. This is especially true of very soft fruit like cherries and peaches, which when fully ripe are so soft as to be damaged by the act of picking itself. Further, since many types of fruit continue to ripen off the tree, unless they were to be processed quickly, some would become overripe before they could be utilised if picked at peak ripeness.

From a technological point of view, fruit characterisation by species and varieties is performed on the basis of physical as well chemical properties: shape, size, texture, flavour, colour/pigmentation, dry matter content (soluble solids content), pectic substances, acidity, vitamins, etc. These properties are directly correlated with fruit utilisation. The following table shows which of the above mentioned properties have a major impact on the finished products obtained by fruit processing.

Table: *Optimal use of fruits as a function of their properties*

Processed Finished Products	***Organoleptic (Sensory) Properties***				***Chemical Composition***		
	Shape	***Texture***	***Flavour***	***Taste***	***Acidity***	***Sugars***	***Pectic Sust.***
Dried Fruits	++	++		++		++	
Fruit Juices			++	++	++		
Marmalade			++	++			++
Jams	++	++	++	++			
Jellies	++	++	++	++			++
Fruit Paste				++		++	

When to Pick

The proper time to pick fruit depends upon several factors; these include variety, location, weather, ease of removal from the tree (which change with time), and purpose to which the fruit will be put.

For example, oranges change with respect to both sugar and acid as they ripen on the tree; sugar increases and acid decreases. The ratio of sugar to acid determines the taste and acceptability of the fruit and the juice. For this reasons, in some countries there are laws that prohibit picking until a certain sugar-acid ratio has been reached.

In the case of much fruit to be canned, on the other hand, fruit is picked before it is fully ripe for eating since canning will further soften the fruit.

Quality Measurements

Many quality measurements can be made before a fruit crop is picked in order to determine if proper maturity or degree of ripeness has developed.

Colour may be measured with instruments or by comparing the colour of fruit on the tree with standard picture charts.

Texture may be measured by compression by hand or by simple type of plungers.

As fruit mature on the tree its concentration of juice solids, which are mostly sugars, changes. The concentration of soluble solids in the juice can be estimated with a refractometer or a hydrometer. The refractometer measures the ability of a solution to bend or refract a light beam which is proportional to the solution's concentration. A hydrometer is a weighted spindle with a graduated neck which floats in the juice at a height related to the juice density.

The acid content of fruit changes with maturity and affects flavour. Acid concentration can be measured by a simple chemical titration on the fruit juice. But for many fruits the tartness and flavour are really affected by the ratio of sugar to acid.

Percentage of soluble solids, which are largely sugars, is generally expressed in degrees Brix, which relates specific gravity of a solution to an equivalent concentration of pure sucrose.

In describing the taste of tartness of several fruits and fruit juices, the term "sugar to acid ratio" or "Brix to acid ratio" are commonly used. The higher the Brix the greater the sugar concentration in the juice; the higher the "Brix to acid ratio" the sweeter and lees tart is the juice.

Harvesting and Preprocessing

Harvesting

The above and other measurements, plus experience, indicate when fruit is ready for harvesting and subsequent processing.

A large amount of the harvesting of most fruit crops is still done by hand; this labour may represent about half of the cost of growing the fruit. Therefore, mechanical harvesting is currently one of the most active fields of research for the agricultural engineer, but also requires geneticists to breed fruit of nearly equal size, that matures uniformly and that is resistant to mechanical damage.

A correct manual harvesting includes some simple but essential rules:

- the fruit should be picked by hand and placed carefully in the harvesting basket; all future handling has to be performed carefully in order to avoid any mechanical damage;
- the harvesting basket and the hands of the harvester should be clean;
- the fruit should be picked when it is ready to be able to be processed into a quality product depending on the treatment which it will undergo.

It is worth emphasising the fact that the proximity of the processing centre to the source of supply for fresh raw materials presents major advantages; some are as follows:

- possibility to pick at the best suitable moment;
- reduction of losses by handling/transportation;
- minimises raw material transport costs;
- possibility to use simpler/cheaper receptacles for raw material transport.

Once it has left the tree, the organoleptic properties, nutritional value, safety and aesthetic appeal of the fruit deteriorates in varying degrees. The major causes of deterioration include the following:

a. growth and activity of micro-organisms;
b. activities of the natural food enzymes;
c. insects, parasites and rodents;
d. temperature, both heat and cold;
e. moisture and dryness;
f. air and in particular oxygen;
g. light and
h. time.

Reception - quality and Quantity

Fruit reception at the processing centre is performed mainly for following purposes:

- checking of sanitary and freshness status;
- control of varieties and fruit wholeness;
- evaluation maturity degree;
- collection of data about quantities received in connection to the source of supply: outside growers/farmers, own farm.

Variety control is needed in order to identify that the fruit belongs to an accepted variety as not all are suitable for different technological processes.

Fruit maturity degree is significant as industrial maturity is required for some processing/preservation methods while for others there is the need for an edible maturity when the fruit has full taste and flavour.

Special attention is given to size, appearance and uniformity of fruit to be processed, mainly in the form of fruit preserved with sugar using whole/half fruits ("with fruit pieces").

Some laboratory control is also needed, even if it not easy to precisely establish the technological qualities of fruit because of the absence of enough reliable rapid analytical methods able to show eventual deterioration. The only reliable method for evaluating the quality is the combination of data obtained through organoleptic/taste controls and by simple analytical checks which are possible to perform in a small laboratory: percentage of soluble solids by refractometer, consistency/texture measured with simple penetrometers, etc.

Temporary Storage Before Processing

This step has to be as short as possible in order to avoid flavour losses, texture modification, weight losses and other deterioration that can take place over this period.

Some basic rules for this step are as follows:

- keep products in the shade, without any possible direct contact with sunlight;
- avoid dust as much as possible;
- avoid excessive heat;
- avoid any possible contamination;
- store in a place protected from possible attack by rodents, insects, etc.

Cold storage is always highly preferred to ambient temperature. For this reason a very good manufacturing practice is to use a cool room for each processing centre; this is very useful for small and medium processing units as well.

Raw Material Control - Fresh Fruits and Vegetables at Reception:

1. Checks at each delivery/raw material lot
   - Colour
   - Texture
   - Taste
   - Flavour
   - Appearance
   - Refractometric extract
   - Number per kg
   - Variety
   - Sanitary evaluation
2. Checks at each ten lots (for the same raw material)
   - Density
   - Water content: oven method
   - Total sugars, reducing sugars
   - Total acidity
3. Audits - every six months - on five different lots
   - Ascorbic acid
   - Mineral substances
   - Tannic substances
   - Pectic substances

The type of analysis for audits will be adapted to the specific fruits and vegetables that are received/ processed.

An excellent indication of a good temporary storage is the limited weight loss before processing, which has to be below 1.0%-1.2%.

Washing

Harvested fruit is washed to remove soil, micro-organisms and pesticide residues.

Fruit washing is a mandatory processing step; it would be wise to eliminate spoiled fruit before washing in order to avoid the pollution of washing tools and/or equipment and the contamination of fruit during washing.

Washing efficiency can me gauged by the total number of micro-organisms present on fruit surface before and after washing - best result are when there is a six fold reduction. The water from the final wash should be free from moulds and yeast; a small quantity of bacteria is acceptable.

Fruit washing can be carried out by immersion, by spray/ showers or by combination of these two processes which is generally the best solution: pre-washing and washing.

Some usual practices in fruit washing are:

- addition of detergents or 1.5% HCl solution in washing water to remove traces of insect-fungicides;
- use of warm water (about 50°C) in the pre-washing phase;
- higher water pressure in spray/shower washers.

Washing must be done before the fruit is cut in order to avoid losing high nutritive value soluble substances (vitamins, minerals, sugars, etc.).

Sorting

Fruit sorting covers two main separate processing operations:

a. removal of damaged fruit and any foreign bodies (which might have been left behind after washing);

b. qualitative sorting based on organoleptic criteria and maturity stage.

Mechanical sorting for size is usually not done at the preliminary stage. The most important initial sorting is for variety and maturity.

However, for some fruit and in special processing technologies it is advisable to proceed to a manual dimensional sorting (grading).

Trimming and Peeling (Skin Removal)

This processing step aims at removing the parts of the fruit which are either not edible or difficult to digest especially the skin.

Up to now the industrial peeling of fruit and vegetables was performed by three procedures:

a. mechanically;
b. by using water steam;
c. chemically; this method consists in treating fruit and vegetables by dipping them in a caustic soda solution at a temperature of 90 to 100° C; the concentration of this solution as well as the dipping or immersion time varying according to each specific case.

Cutting

This step is performed according to the specific requirements of the fruit processing technology.

Heat Blanching

Fruit is not usually heat blanched because of the damage from the heat and the associated sogginess and juice loss after thawing. Instead, chemicals are commonly used without heat to inactivate the oxidative enzymes or to act as antioxidants and they are combined with other treatments.

Ascorbic/citric acid Dip

Ascorbic acid or vitamin C minimises fruit oxidation primarily by acting as an antioxidant and itself becoming oxidised in preference to catechol-tannin compounds. Ascorbic acid is frequently used by being dissolved in water, sugar syrup or in citric acid solutions.

It has been found that increased acidity also helps retard oxidative colour changes and so ascorbic acid plus citric acid may be used together. Citric acid further reacts with (chelates) metal ions thus removing these catalysts of oxidation from the system.

Sulphur Dioxide Treatment

Sulphur dioxide may function in several ways:

- sulphur dioxide is an enzyme poison against common oxidising enzymes;
- it also has antioxidant properties; i.e., it is an oxygen acceptor (as is ascorbic acid);

- further SO2 minimises non enzymatic Maillard type browning by reacting with aldehyde groups of sugars so that they are no longer free to combine with amino acids;
- sulphur dioxide also interferes with microbial growth.

In many fruit processing pre-treatments two factors must be considered:

a. sulphur dioxide must be given time to penetrate the fruit tissues;
b. SO2 must not be used in excess because it has a characteristic unpleasant taste and odour, and international food laws limit the SO2 content of fruit products, especially of those which are consumer oriented (e.g. except semi-processed products oriented to further industrial utilisation).

Commonly a 0.25 % solution (except for semi-processed fruit products which are industry oriented and use a 6% solution) of SO2 or its SO2 equivalent in the form of solutions of sodium sulphite, sodium bisulphite or sodium/potassium metabisulphite are used.

Fruit slices are dipped in the solution for about two to three minutes and then removed so as not to absorb too much SO2. Then the slices are allowed to stand for about one to two hours so that the SO2 may penetrate throughout the tissues before processing.

Sulphur dioxide is also used in fruit juice production to minimise oxidative changes where relatively low heat treatment is employed so as not to damage delicate juice flavour.

Dry sulphuring is the technological step where fruit is exposed to fumes of SO2 from burning sulphur or from compressed gas cylinders; this treatment could be used in the preparation of fruits (and some vegetables) prior to drying / dehydration.

Sugar Syrup

Sugar syrup addition is one of the oldest methods of minimising oxidation. It was used long before the causative reactions were understood and remains today a common practice for this purpose.

Sugar syrup minimises oxidation by coating the fruit and thereby preventing contact with atmospheric oxygen.

Sugar syrup also offers some protection against loss of volatile fruit esters and it contributes sweet taste to otherwise tart fruits.

It is common today to dissolve ascorbic acid and citric acid in the sugar syrup for added effect or to include sugar syrup after an SO2 treatment.

Fresh Fruit Storage

Some fruit species and specially apples and pears can be stored in fresh state during cold season in some countries' climatic conditions.

Fruit for fresh storage have to be autumn or winter varieties and be harvested before they are fully mature. This fruit also has to be sound and without any bruising; control and sorting by quality are mandatory operations.

Sorting has to be carried out according to size and weight and also by appearance; fruit which is not up to standard for storage will be used for semi-processed product manufacturing which will be submitted further to industrial processing.

Harvested fruit has to be transported as soon as possible to storage areas. Leaving fruit in bulk in order to generate transpiration is a bad practice as this reduces storage time and accelerates maturation processes during storage.

In order to store large quantities of fruit, silos have to be built.

Vegetable Specific Processing Technologies

In culinary terms, a vegetable is an edible plant or its part, intended for cooking or eating raw. In biological terms, "vegetable" designates members of the plant kingdom.

The non-biological definition of a vegetable is largely based on culinary and cultural tradition. Apart from vegetables, other main types of plant food are fruits, grains and nuts. Vegetables are most often consumed as salads or cooked in savory or salty dishes, while culinary fruits are usually sweet and used for desserts, but it is not the universal rule. Therefore, the division is somewhat arbitrary, based on cultural views. For example, some people consider mushrooms to be vegetables even though they are not biologically plants, while others consider them a separate food category; some cultures group potatoes with cereal products such as noodles or rice, while most English speakers would consider them vegetables.

Some vegetables can be consumed raw, some may be eaten cooked, and some must be cooked to destroy certain natural toxins or microbes in order to be edible, such as eggplant, unripe potatoes, daylily, winter melon, fiddlehead fern, and most kinds of legume/beans (such as common beans). A number of processed food items available on the market contain vegetable ingredients and can be referred to as "vegetable derived" products. These products may or may not maintain the nutritional integrity of the vegetable used to produce them.

The word "vegetable" was first recorded in English in the 15th century, and originally applied to any plant. This is still the sense of the adjective "vegetable" in biological context. In 1767, the meaning of the term "vegetable" was specified to mean "plant cultivated for food, edible herb or root." The year 1955 noted the first use of the shortened, slang term "veggie".

As an adjective, the word vegetable is used in scientific and technical contexts with a different and much broader meaning, namely of "related to plants" in general, edible or not — as in *vegetable matter*, *vegetable kingdom*, *vegetable origin*, etc. The meaning of "vegetable" as "plant grown for food" was not established until the 18th century.

Terminology

There are at least four definitions relating to fruits and vegetables:

- Fruit (botany): the ovary of a flowering plant (sometimes including accessory structures),
- Fruit (culinary): any edible part of a plant with a sweet flavour,
- Vegetable (culinary): any edible part of a plant with a savory flavour.
- Vegetable (legal): commodities that are taxed as vegetables in a particular jurisdiction

In everyday, grocery-store, culinary language, the words "fruit" and "vegetable" are mutually exclusive; plant products that are called fruit are hardly ever classified as vegetables, and vice-versa. The word "fruit" has a precise botanical meaning (a part that developed from the ovary of a flowering plant), which is considerably different from its culinary meaning, and includes many poisonous fruits. While peaches, plums, and oranges are "fruit" in both senses, many items commonly called "vegetables" — such as eggplants, bell peppers, and tomatoes — are botanically fruits, while the cereals (grains) are both a fruit and a vegetable, as well as some spices like black pepper and chili peppers. The question of whether the tomato is a fruit or a vegetable found its way into the United States Supreme Court in 1893. The court ruled unanimously in *Nix v. Hedden* that a tomato is correctly identified as, and thus taxed as, a vegetable, for the purposes of the Tariff of 1883 on imported produce. The court did acknowledge, however, that, botanically speaking, a tomato is a fruit.

Languages other than English often have categories that can be identified with the common English meanings of "fruit" and "vegetable", but their precise meaning often depends on local culinary traditions. For example, in Brazil the avocado is traditionally consumed with sugar as a dessert or in milkshakes, and hence it is regarded as a culinary fruit; whereas in other countries (including Mexico and the United States) it is used in salads and dips, and hence considered to be a vegetable.

Examples of Different Parts of Plants used as Vegetables

Figure: *Farmers' market showing vegetables for sale near the Potala palace in Lhasa, Tibet.*

The list of food items called "vegetable" is quite long, and includes many different parts of plants:

Flower Bud: broccoli, cauliflower, globe artichokes, capers

Leaves: kale, collard greens, spinach, arugula, beet greens, bok choy, chard, choi sum, turnip greens, endive, lettuce, mustard greens, watercress, garlic chives, gai lan

Leaf Sheaths: leeks

Buds: Brussels sprouts

Stem: Kohlrabi, galangal, and ginger

Stems of Leaves: celery, rhubarb, cardoon, Chinese celery

Stem Shoots: asparagus, bamboo shoots

Tubers: potatoes, Jerusalem artichokes, sweet potatoes, taro, and yams

Whole-plant Sprouts: soybean, mung beans, urad, and alfalfa

Roots: carrots, parsnips, beets, radishes, rutabagas, turnips, and burdocks

Bulbs: onions, shallots, garlic

Fruits in the Botanical Sense, but Used as Vegetables: tomatoes, cucumbers, squash, zucchinis, pumpkins, peppers, eggplant, tomatillos, chayote, okra, breadfruit, avocado, pods, seeds such as corn, green beans and snow peas.

Nutrition

Vegetables are eaten in a variety of ways, as part of main meals and as snacks. The nutritional content of vegetables varies considerably, though generally they contain little protein or fat, and varying proportions of vitamins such as Vitamin A, Vitamin K and Vitamin B6, provitamins, dietary minerals and carbohydrates. Vegetables contain a great variety of other phytochemicals, some of which have been claimed to have antioxidant, antibacterial, antifungal, antiviral and anticarcinogenic properties. Some vegetables also contain fibre, important for gastrointestinal function. Vegetables contain important nutrients necessary for healthy hair and skin as well. A person who refrains from dairy and meat products, and eats only plants (including vegetables) is known as a vegan.

However, vegetables often also contain toxins and antinutrients such as α-solanine, α-chaconine, enzyme inhibitors (of cholinesterase, protease, amylase, etc.), cyanide and cyanide precursors, oxalic acid, and more. Depending on the concentration, such compounds may reduce the edibility, nutritional value, and health benefits of dietary vegetables. Cooking and/or other processing may be necessary to eliminate or reduce them.

Diets containing recommended amounts of fruits and vegetables may help lower the risk of heart diseases and type 2 diabetes. These diets may also protect against some cancers and decrease bone loss. The potassium provided by both fruits and vegetables may help prevent the formation of kidney stones.

Dietary Recommendations

The USDA Dietary Guidelines for Americans recommends consuming 3 to 5 servings of vegetables daily. This recommendation can vary based on age and gender, and is determined based upon standard portion sizes typically consumed, as well as general nutritional content. For most vegetables, one serving is equal to 1/2 cup and can be eaten raw or cooked. For leafy greens, such as lettuce and spinach, a single serving is typically 1 cup. Serving sizes for vegetable-derived products have not been definitively determined, but usually follow the 1/2 cup standard. Examples of vegetable-derived products subject to this standard are ketchup, pizza sauce, and tomato paste. Currently, there is no specific standard for measuring a vegetable serving in regards to its nutrient content, since different vegetables contain a wide variety of nutrients.

International dietary guidelines are similar to the ones established by the USDA. Japan, for example, recommends the consumption of 5 to 6 servings of vegetables daily. French dietary guidelines follow similar guidelines and set the daily goal at 5 servings.

Colour Pigments

The green colour of leaf vegetables is due to the presence of the green pigment chlorophyll. Chlorophyll is affected by the pH, and it changes to olive green in acid conditions, and to bright green in alkaline conditions. Some of the acids are released in steam during cooking, particularly if cooked without a cover.

The yellow/orange colours of fruits and vegetables are due to the presence of carotenoids, which are also affected by normal cooking processes or changes in pH.

The red/blue colouring of some fruits and vegetables (e.g., blackberries and red cabbage) are due to anthocyanins, which are sensitive to changes in pH. When the pH is neutral, the pigments are purple, when acidic, red, and when alkaline, blue. These pigments are quite water-soluble. This property can be used in rudimentary testing of pH.

Commercial Cultivation

Method of Vegetable Production: Most vegetables are produced using traditional farming practices. However, the yield of vegetables from organic farming is growing.

Origin Countries: Of all the world's nations, China is the leading cultivator of vegetables, with top productions in potato, onions, cabbage, lettuce, tomatoes and broccoli.

Domestic Cultivation: Amateur vegetable growing has a long history dating back to the establishment of agriculture and is a substantial business in itself, with one major online UK seed supplier currently offering well over 1200 cultivars of seeds, plants and potato tubers. They report that sale of vegetable seeds now outstrips that of flower seeds.

Plants are usually divided into the following categories:-

- Root vegetables (carrot, parsnip, beetroot, turnip)
- Leafy green vegetables (cabbage, broccoli, Brussels sprout, kale, spinach, cauliflower)
- Alliums (onion, leek, garlic, shallot)
- Potatoes
- Legumes (peas and beans)
- Pumpkin family (cucumber, courgette, marrow, squash, pumpkin, gourd)
- Tomatoes
- Salad crops (lettuce, radish, rocket, celery, cress, bean sprouts)
- Other vegetables (pepper, sweet potato, corn, artichoke, asparagus, mushroom)

Each group has its own cultivation needs, but the overall requirements for the vast majority of vegetable plants are:

- deep, rich soil with a neutral or slightly alkaline composition
- plenty of sunshine
- plenty of water
- regular feeding
- regular weeding
- protection against pests such as slugs, aphids and caterpillars

The substantial literature and other media on the subject advise deep digging, annual top dressing with manure or home-made compost, and crop rotation. Other issues are also discussed at length, including:-

- raised beds
- successional sowing
- garden design
- allotments
- raising plants for competition
- comparative evaluation of different varieties
- organic gardening

Increasingly, small plants (plug plants) are offered for sale in spring and summer.

Storage

Proper post-harvest storage aimed at extending and ensuring shelf life is best effected by efficient cold chain application. All vegetables benefit from proper post harvest care.

Many root and non-root vegetables that grow underground can be stored through winter in a root cellar or other similarly cool, dark, and dry place to prevent the growth of mold, greening and sprouting. Care should be taken in understanding the properties and vulnerabilities of the particular roots to be stored. These vegetables can last through to early spring and be nearly as nutritious as when fresh.

During storage, leafy vegetables lose moisture, and the vitamin C in them degrades rapidly. They should be stored for as short a time as possible in a cool place, in a sealed container or a plastic bag.

5

Processing Technologies

Vegetables Varieties

Vegetable processors must appreciate the substantial differences that varieties of a given vegetable will possess. In addition to variety and genetic strain differences with respect to weather, insect and disease resistance, varieties of a given vegetable will differ in size, shape, time of maturity, and resistance to physical damage.

Varietal differences then further extend into warehouse storage stability, and suitability for such processing methods as canning, freezing, pickling or drying. A variety of peas that is suitable for canning may be quite unsatisfactory for freezing and varieties of potatoes that are preferred for freezing may be less satisfactory for drying or potato chip manufacture.

This should be expected since different varieties of a given vegetable will vary somewhat in chemical composition, cellular structure and biological activity of their enzyme system.

Harvesting and Pre-processing

When vegetables are maturing in the field they are changing from day to day. There is a time when the vegetable will be at peak quality from the stand-point of colour, texture and flavour.

This peak quality is quick in passing and may last only a day. Harvesting and processing of several vegetables, including tomatoes, corn and peas are rigidly scheduled to capture this peak quality.

After the vegetable is harvested it may quickly pass beyond the peak quality condition. This is independent of microbiological spoilage; these main deteriorations are related to:

a) loss of sugars due to their consumption during respiration or their conversion to starch; losses are slower under refrigeration but there is still a great change in vegetable sweetness and freshness of flavour within 2 or 3 days;

b) production of heat when large stockpiles of vegetables are transported or held prior to processing.

At room temperature some vegetables will liberate heat at a rate of 127,000 kJ/ton/day; this is enough for each ton of vegetables to melt 363 kg of ice per day. Since the heat further deteriorates the vegetables and speeds micro-organisms growth, the harvested vegetables must be cooled if not processed immediately.

But cooling only slows down the rate of deterioration, it does not prevent it, and vegetables differ in their resistance to cold storage. Each vegetable has its optimum cold storage temperature which may be between about 0-100 C (32-500 F).

c) the continual loss of water by harvested vegetables due to transpiration, respiration and physical drying of cut surfaces results in wilting of leafy vegetables, loss of plumpness of fleshy vegetables and loss of weight of both.

Moisture loss cannot be completely and effectively prevented by hermetic packaging. This was tried with plastic bags for fresh vegetables in supermarkets but the bags became moisture fogged, and deterioration of certain vegetables was accelerated because of buildup of CO2 and decrease of oxygen in the package. It therefore is common to perforate such bags to prevent these defects as well as to minimise high humidity in the package which would encourage microbial growth.

Shippers of fresh vegetables and vegetable processors, whether they can, freeze, dehydrate, or manufacture soups or ketchup, appreciate the instability and perishability of vegetables and so do everything they can to minimise delays in processing of the fresh product. In many processing plants it is common practice to process vegetables immediately as they are received from the field.

To ensure a steady supply of top quality produce during the harvesting period the large food processors will employ trained field men; they will advise on growing practices and on spacing of plantings so that vegetables will mature and can be harvested in rhythm with the processing plant capabilities. This minimises stockpiling and need for storage.

Cooling of vegetables in the field is common practice in some areas. Liquid nitrogen-cooled trucks may next provide transportation of fresh produce to the processing plant or directly to market.

Upon arrival of vegetables at the processing centre the usual operations of cleaning, grading, peeling, cutting and the like are performed using a moderate amount of equipment but a good deal of hand labour also still remains.

Reception

This covers qualitative and quantitative control of delivered vegetables. The organoleptic control and the evaluation of the sanitary state, even if they are very important steps in vegetables' characteristics assessment, cannot establish their technological value.

On the other hand, laboratory controls do not precisely establish their technological properties because of the difficulty in putting into showing some deterioration when using rapid control methods.

One correct method of vegetable quality appraisal is their overall evaluation based on the whole complex of data that can be obtained by combining an extensive organoleptic evaluation with simple analysis that can be performed rapidly in plant laboratory. These analysis can be:

a. refractometric extract (tomatoes, fruit, etc.);

b. specific weight (potatoes, peas, etc.);

c. consistency (measured with tenderometers, penetrometers, etc.);

d. boiling tests, etc.

Temporary Storage

This step should be as short as possible and better completely eliminated. Vegetables can be stored in:

a. simple stores, without artificial cooling;

b. in refrigerated stores; or, in some cases,

c. in silos (potatoes, etc.).

Simple stores should be covered, fairly cool, dry and well ventilated but without forced air circulation which can induce significant losses in weight through intensive water evaporation; air relative humidity should be at about 70-80%.

Refrigerated storage is always preferable and in all cases a processing centre needs a cold room for this purpose, adapted in volume I capacity to the types and quantities of vegetables (and fruits) that are further processed.

Washing

Washing is used not only to remove field soil and surface micro-organisms but also to remove fungicides, insecticides and other pesticides, since there are laws specifying maximum levels of these materials that may be retained on the vegetable; and in most cases the allowable residual level is virtually zero. Washing water contains detergents or other sanitisers that can essentially completely remove these residues.

The washing equipment, like all equipment subsequently used, will depend upon the size, shape and fragility of the particular kind of vegetable:

- flotation cleaner for peas and other small vegetables;
- rotary washer in which vegetables are tumbled while they are sprayed with jets of water; this type of washer should not be used to clean fragile vegetables;

Skin Removal/peeling

Some vegetables require skin removal. This can be done in various ways.

Mechanical: This type of operation is performed with various types of equipment which depend upon the result expected and the characteristics of the fruit and vegetables, for example:

i. a machine with abrasion device (potatoes, root vegetables);
ii. equipment with knives (apples, pears, potatoes, etc.);
iii. equipment with rotating sieve drums (root vegetables). Sometimes this operation is simultaneous with washing (potatoes, carrots, etc.) or preceded by blanching (carrots).

Chemical: Skins can be softened from the underlying tissues by submerging vegetables in hot alkali solution. Lye may be used at a concentration of about 0.5-3%, at about 93° C (2000 F) for a short time period (0.5-3 min). The vegetables with loosened skins are then conveyed under high velocity jets of water which wash away the skins and residual lye.

In order to avoid enzymatic browning, this chemical peeling is followed by a short boiling in water or an immersion in diluted citric acid solutions.

It is more difficult to peel potatoes with this method because it is necessary to dissolve the cutin and this requires more concentrated lye solutions, up to 10%.

Thermal: Wet heat (steam). Other vegetables with thick skins such as beets, potatoes, carrots and sweet potatoes may be peeled with steam under pressure (about 10 at) as they pass through cylindrical vessels. This softens the skin and the underlying tissue. When the pressure is suddenly released, steam under the skin expands and causes the skin to puff and crack. The skins are then washed away with jets of water at high pressure (up to 12 at).

Dry heat (flame). Other vegetables such as onions and peppers are best skinned by exposing them to direct flame (about 1 min at 1000° C) or to hot gases in rotary tube flame peelers. Here too, heat causes steam to develop under skins and puff them so that they can be washed away with water.

Manual peeling only use when the other methods are impossible or sometimes as a completion of the other three ways. Average losses at this step are given in table.

Table: *Losses at vegetable peeling, in %*

	Peeling methods		
Vegetables	***Manual***	***Mechanical***	***Chemical***
Potatoes	15-19	18-28	-
Carrots	13-15	16-18	8-10
Beets	1416	13-15	9-10

Size Reduction

This step is applied according to specific vegetable and processing technology requirements.

Blanching

The special heat treatment to inactivate enzymes is known as blanching. Blanching is not indiscriminate heating. Too little is ineffective, and too much damages the vegetables by excessive cooking, especially where the fresh character of the vegetable is subsequently to be preserved by processing.

This heat treatment is applied according to and depends upon the specificity of vegetables, the objectives that are followed and the subsequent processing / preservation methods. Two of the more heat resistant enzymes important in vegetables are catalase and peroxidase.

If these are destroyed then the other significant enzymes in vegetables also will have been inactivated. The heat treatment to destroy catalase and peroxidase in different vegetables are known, and sensitive chemical tests have been developed to detect the amounts of these enzymes that might survive a blanching treatment.

Because various types of vegetables differ in size, shape, heat conductivity, and the natural levels of their enzymes, blanching treatments have to be established on an experimental basis. As with sterilisation of foods in cans, the larger the food item the longer it takes for heat to reach the centre. Small vegetables may be adequately blanched in boiling water in a minute or two, large vegetables may require several minutes.

Blanching as a unit operation is a short time heating in water at temperatures of 100° C or below. Water blanching may be performed in double bottom kettles, in special baths with conveyor belts or in modern continuous blanching equipment.

In order to reduce losses of hydrosoluble substances (mineral salts, vitamins, sugars, etc.) occurring during water blanching, several methods have been developed:

- temperature setting at 85-95° C instead of 100° C;
- blanching time has to be just sufficient to inactivate enzymes catalase and peroxidase;
- assure elimination of air from tissues.

Table: *Blanching parameters for some vegetables*

Vegetables	***Temperature, °C***	***Time, min.***
Peas	85-90	2-7
Green beans	90-95	2-5
Cauliflower	Boiling	2
Carrots	90	3-5
Peppers	90	3

Steam heat treatment can also be applied instead of water blanching as a preliminary step before freezing or drying, as long as the preservation method is only used for enzyme inactivation and not to modify consistency.

For drying, the vegetables are conveyed directly from steaming equipment to drying installations without cooling. Vegetable steaming is carried out in continuous installations with conveyer belts made from metallic sieves.

Cooling of vegetables after water blanching or steaming is performed in order to avoid excessive softening of the tissues and has to follow immediately after these operations; one exception is the case of vegetables for drying which can be transferred directly to drying equipment without cooling.

Natural cooling is not recommended because is too long and generates significant losses in vitamin C content. Cooling in pre-cooled air (from special installations) is sometimes used for vegetables that will be frozen

Cooling in water can be achieved by sprays or by immersion; in any case the vegetables have to reach a temperature value under 37° C as soon as possible. Too long a cooling time generates supplementary losses in valuable hydrosoluble substances; in order to avoid this, the temperature of the cooling water has to be as low as possible.

Canning

Large quantities of vegetable products are canned. A typical flow sheet for a vegetable canning operation (which also applies to fruit for the most part) covers some food process unit operations performed in sequence: harvesting; receiving; washing; grading; heat blanching; peeling and coring; can filling; removal of air under vacuum; sealing/closing, retorting/heat treatment; cooling; labelling and packing. The vegetable may be canned whole, diced, puréed, as juice and so on.

Online Simplified Methods for Enzyme Activity Check

Peroxidase test

a) Solutions. In order to check the peroxidase activity two solutions have to be prepared:
   - 1% guaiacol in alcohol solution (1 g guaiacol is dissolved in about 50 cm^3 of 96% ethylic alcohol and then this preparation is brought to 100 cm° with the same solvent);
   - peroxide solution 0.3% (1 cm^3 perhydrol is brought to 100 cm^3 with distilled water.

b) Sampling. From various parts of the material samples are taken (about 20-30 pieces, etc.); the material is then crushed in a laboratory bowl in order to obtain an average sample.

c) Check. Prom the average sample, 10-20 g of material is taken in a medium capacity test tube; on this sample are poured: 20 cm^3 distilled water; 1 cm^3 of 1% guaiacol solution; 1.6 cm^3 of peroxide solution.

The contents of the test tube is shaken well. The gradual appearance of a weak pink colour indicates an incomplete peroxidase inactivation - reaction slightly positive. If there are no tissue colour modifications after 5 minutes, the reaction is negative and the enzymes have been inactivated.

As an orientative check it is also possible to simply pour a few drops of 1% guaiacol solution and 0.3% peroxide solution directly on blanched and crushed vegetables. A rapid and intensive brown-reddish tissue colouring indicates a high peroxidase activity (positive reaction).

Catalase Test

In order to identify the catalase enzyme activity, 2 g of dehydrated vegetables are well crushed and mixed with about 20 cm^3 of distilled water. After 15 min softening, 0.5 cm^3 of a 0.5% or 1% peroxide solution is poured on prepared vegetables. In the presence of catalase, a strong oxygen generation is observed for about 2-3 minutes.

These tests are of a paramount importance in order to determine the vegetable blanching treatments (temperature and time); incomplete enzyme inactivation has a negative effect on finished product quality.

For cabbage catalase inactivation by blanching is sufficient; blanching further to peroxidase inactivation would have negative effects on product quality and even complete browning.

For all other vegetables and for potatoes, both tests MUST be negative, for catalase and for peroxidase.

Coffee Processing

Coffee is a brewed beverage prepared from the roasted seeds of several species of an evergreen shrub of the genus *Coffea.* The two most common sources of coffee beans are the highly regarded *Coffea arabica*, and the "robusta" form of the hardier *Coffea canephora.* The latter is resistant to the coffee leaf rust (*Hemileia vastatrix*), but has a more bitter taste. Coffee plants are cultivated in over 70 countries, primarily in equatorial Latin America, Southeast Asia, Maldives, and Africa. Once ripe, coffee "berries" are picked, processed, and dried to yield the seeds inside. The seeds are then roasted to varying degrees, depending on the desired flavour, before being ground and brewed to create coffee.

Coffee is slightly acidic (pH 5.0–5.1) and can have a stimulating effect on humans because of its caffeine content. It is one of the most popular drinks in the world. It can be prepared and presented in a variety of ways. Many studies have examined the health effects of coffee, and whether the overall effect of coffee consumption is positive or negative has been widely disputed. The majority of recent research suggests that moderate coffee consumption is benign or mildly beneficial in healthy adults. However, coffee can worsen the symptoms of some conditions such as anxiety, largely due to the caffeine and diterpenes it contains.

Coffee cultivation first took place in southern Arabia; the earliest credible evidence of coffee-drinking appears in the middle of the 15th century in the Sufi shrines of Yemen. In East Africa and Yemen, coffee was used in native religious ceremonies that were in competition with the Christian Church. As a result, the Ethiopian Church banned its secular consumption until the reign of Emperor Menelik II of Ethiopia. The beverage was also banned in Ottoman Turkey during the 17th century for political reasons and was associated with rebellious political activities in Europe.

An important export commodity, coffee was the top agricultural export for twelve countries in 2004, and it was the world's seventh-largest legal agricultural export by value in 2005. Green (unroasted) coffee is one of the most traded agricultural commodities in the world. Some controversy is associated with coffee cultivation and its impact on the environment. Consequently, organic coffee is an expanding market.

History

***Figure:** Over the door of a Leipzig coffeeshop is a sculptural representation of a man in Turkish dress, receiving a cup of coffee from a boy*

Legendary Accounts

According to legend, ancestors of today's Oromo people were believed to have been the first to recognize the energizing effect of the coffee plant, though no direct evidence has been found indicating where in Africa coffee grew or who among the native populations might have used it as a stimulant or even known about it, earlier than the 17th century. The story of Kaldi, the 9th-century Ethiopian goatherder who supposedly discovered coffee when his goats behaved strangely after eating from a coffee plant, did not appear in writing until 1671 and is probably apocryphal. The original domesticated coffee plant is said to have been from Harar.

Other accounts attribute the discovery of coffee to Sheik Omar. According to the ancient chronicle (preserved in the Abd-Al-Kadir manuscript), Omar, who was known for his ability to cure the sick through prayer, was once exiled from Mocha, Yemen to a desert cave near Ousab. Starving, Omar chewed berries from nearby shrubbery, but found them to be bitter. He tried roasting the seeds to improve the flavour, but they became hard. He then tried boiling them to soften the seed, which resulted in a fragrant brown liquid. Upon drinking the liquid Omar was revitalized and sustained for days. As stories of this "miracle drug" reached Mocha, Omar was asked to return and was made a saint. From Ethiopia, the beverage was introduced into the Arab world through Egypt and Yemen.

Historical Transmission

The earliest credible evidence of either coffee drinking or knowledge of the coffee tree appears in the middle of the 15th century, in the Sufi monasteries around Mocha in Yemen. It was here in Arabia that coffee seeds were first roasted and brewed, in a similar way to how it is now prepared. By the 16th century, it had reached the rest of the Middle East, Persia, Turkey, and northern Africa. Coffee seeds were first exported from Ethiopia to Yemen. Yemeni traders brought coffee back to their homeland and began to cultivate the seed. The first coffee smuggled out of the Middle East was by Sufi Baba Budan from Yemen to India in 1670. Before then, all exported coffee was boiled or otherwise sterilised. Portraits of Baba Budan depict him as having smuggled seven coffee seeds by strapping them to his chest. The first plants grown from these smuggled seeds were planted in Mysore. Coffee then spread to Italy, and to the rest of Europe, to Indonesia, and to the Americas.

From the Middle East, coffee spread to Italy. The thriving trade between Venice and North Africa, Egypt, and the Middle East brought many goods, including coffee, to the Venetian port. From Venice, it was introduced to the rest of Europe. Coffee became more widely accepted after it was deemed a Christian beverage by Pope Clement VIII in 1600, despite appeals to ban the "Muslim drink." The first European coffee house opened in Italy in 1645.

Figure: *A Coffee can from the first half of the 20th century. From the Museo del Objeto del Objeto collection.*

The Dutch East India Company was the first to import coffee on a large scale. The Dutch later grew the crop in Java and Ceylon. The first exports of Indonesian coffee from Java to the Netherlands occurred in 1711.

Through the efforts of the British East India Company, coffee became popular in England as well. Oxford's Queen's Lane Coffee House, established in 1654, is still in existence today. Coffee was introduced in France in 1657, and in Austria and Poland after the 1683 Battle of Vienna, when coffee was captured from supplies of the defeated Turks.

When coffee reached North America during the Colonial period, it was initially not as successful as it had been in Europe as alcoholic beverages remained more popular. During the Revolutionary War, the demand for coffee increased so much that dealers had to hoard their scarce supplies and raise prices dramatically; this was also due to the reduced availability of tea from British merchants, and a general resolution among many Americans to avoid drinking it due to the Boston Tea Party.

After the War of 1812, during which Britain temporarily cut off access to tea imports, the Americans' taste for coffee grew. Coffee consumption declined in England, giving way to tea during the 18th century. The latter beverage was simpler to make, and had become cheaper with the British conquest of India and the tea industry there. During the Age of Sail, seamen aboard ships of the British Royal Navy made substitute coffee by dissolving burnt bread in hot water.

The Frenchman Gabriel de Clieu brought a coffee plant to the French territory of Martinique in the Caribbean, from which much of the world's cultivated arabica coffee is descended. Coffee thrived in the climate and was conveyed across the Americas. The territory of Santo Domingo (now Haiti) saw coffee cultivated from 1734, and by 1788 it supplied half the world's coffee. The conditions that the slaves worked in on coffee plantations were a factor in the soon to follow Haitian Revolution. The coffee industry never fully recovered there.

Meanwhile, coffee had been introduced to Brazil in 1727, although its cultivation did not gather momentum until independence in 1822. After this time, massive tracts of rainforest were cleared first from the vicinity of Rio and later São Paulo for coffee plantations. Cultivation was taken up by many countries in Central America in the latter half of the 19th century, and almost all involved the large-scale displacement and exploitation of the indigenous people. Harsh conditions led to many uprisings, coups and bloody suppression of peasants. The notable exception was Costa Rica, where lack of ready labour prevented the formation of large farms. Smaller farms and more egalitarian conditions ameliorated unrest over the 19th and 20th centuries.

Coffee has become a vital cash crop for many developing countries. Over one hundred million people in developing countries have become dependent on coffee as their primary source of income. It has become the primary export and backbone for African countries like Uganda, Burundi, Rwanda, and Ethiopia, as well as many Central American countries.

Biology

Several species of shrub of the genus *Coffea* produce the berries from which coffee is extracted. The two main species commercially cultivated are *Coffea canephora* (predominantly a form known as 'robusta') and *C. arabica*. *C. arabica*, the most highly regarded species, is native to the southwestern highlands of Ethiopia and the Boma Plateau in southeastern Sudan and possibly Mount Marsabit in northern Kenya. *C. canephora* is native to western and central Subsaharan Africa, from Guinea to the Uganda and southern Sudan.

Less popular species are *C. liberica*, *C. stenophylla*, *C. mauritiana*, and *C. racemosa*.

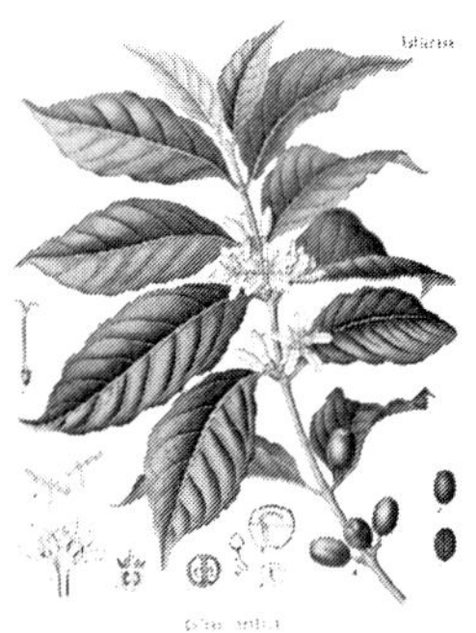

***Figure:** Illustration of* Coffea arabica *plant and seeds*

All coffee plants are classified in the large family Rubiaceae. They are evergreen shrubs or small trees that may grow 5 m (15 ft) tall when unpruned. The leaves are dark green and glossy, usually 10–15 cm (4–6 in) long and 6 cm (2.4 in) wide. The flowers are axillary, and clusters of fragrant white flowers bloom simultaneously and are followed by oval berries of about 1.5 cm (0.6 in). Green when immature, they ripen to yellow, then crimson, before turning black on drying. Each berry usually contains two seeds, but 5–10% of the berries have only one; these are called peaberries. Arabica berries ripen in six to eight months, while robusta take nine to eleven months.

Coffea arabica is predominantly self-pollinating, and as a result the seedlings are generally uniform and vary little from their parents. In contrast, *Coffea canephora*, and *C. liberica* are self-incompatible and require outcrossing. This means that useful forms and hybrids must be propagated vegetatively. Cuttings, grafting, and budding are the usual methods of vegetative propagation. On the other hand, there is great scope for experimentation in search of potential new strains.

Cultivation

The traditional method of planting coffee is to place 20 seeds in each hole at the beginning of the rainy season. This method loses about 50% of the seeds' potential, as about half fail to sprout. A more effective method of growing coffee, used in Brazil, is to raise seedlings in nurseries that are then planted outside at six to twelve months. Coffee is often intercropped with food crops, such as corn, beans, or rice during the first few years of cultivation as farmers become familiar with its requirements.

Of the two main species grown, arabica coffee (from *C. arabica*) is generally more highly regarded than robusta coffee (from *C. canephora*); robusta tends to be bitter and have less flavour but better body than arabica. For these reasons, about three-quarters of coffee cultivated worldwide is *C. arabica*. Robusta strains also contain about 40–50% more caffeine than arabica. For this reason, it is used as an inexpensive substitute for arabica in many commercial coffee blends. Good quality robusta seeds are used in traditional Italian espresso blends to provide a full-bodied taste and a better foam head (known as *crema*).

However, *Coffea canephora* is less susceptible to disease than *C. arabica* and can be cultivated in lower altitudes and warmer climates where *C. arabica* will not thrive. The robusta strain was first collected in 1890 from the Lomani River, a tributary of the Congo River, and was conveyed from Zaire (now the Democratic Republic of Congo) to Brussels to Java around 1900. From Java, further breeding resulted in the establishment of robusta plantations in many countries. In particular, the spread of the devastating coffee leaf rust (*Hemileia vastatrix*), to which *C. arabica* is vulnerable, hastened the uptake of the resistant robusta. Coffee leaf rust is found in virtually all countries that produce coffee.

Over 900 species of insect have been recorded as pests of coffee crops worldwide. Of these, over a third are beetles, and over a quarter are bugs. Some 20 species of nematodes, 9 species of mites, several snails and slugs also attack the crop. Birds and rodents sometimes eat coffee berries but their impact is minor compared to invertebrates. In general, *arabica* is the more sensitive species to invertebrate predation overall. Each part of the coffee plant is assailed by different animals. Nematodes attack the roots, and borer beetles burrow into stems and woody material, the foliage is attacked by over 100 species of larvae (caterpillars) of butterflies and moths.

Mass spraying of insecticides has often proven disastrous, as the predators of the pests are more sensitive than the pests themselves. Instead, integrated pest management has developed, using techniques such as targeted treatment of pest outbreaks, and managing crop environment away from conditions favouring pests. Branches infested with scale are often cut and left on the ground, which promotes scale parasites to not only attack the scale on the fallen branches but in the plant as well.

The 2-mm-long coffee berry borer beetle is the most damaging insect pest to the world's coffee industry, destroying up to 50 percent or more of the coffee berries on plantations in most coffee-producing countries. The adult female beetle nibbles a single tiny hole in a coffee berry and lays 35 to 50 eggs. Inside, the offspring grow, mate, and then emerge from the commercially ruined berry to disperse, repeating the cycle. Pesticides are mostly ineffective because the beetle juveniles are protected inside the berry nurseries, but they are vulnerable to predation by birds when they emerge. When groves of trees are nearby, the American Yellow Warbler, Rufous-capped Warbler and other insectivorous birds have been shown to reduce by 50 percent the number of coffee berry borer beetles in Costa Rica coffee plantations.

World Production

In 2011 Brazil was the world leader in production of green coffee, followed by Vietnam, Indonesia and Colombia. Arabica coffee seeds are cultivated in Latin America, eastern Africa, Arabia, or Asia. Robusta coffee seeds are grown in western and central Africa, throughout southeast Asia, and to some extent in Brazil.

Seeds from different countries or regions can usually be distinguished by differences in flavour, aroma, body, and acidity. These taste characteristics are dependent not only on the coffee's growing region, but also on genetic subspecies (varietals) and processing. Varietals are generally known by the region in which they are grown, such as Colombian, Java and Kona.

Ecological Effects

Originally, coffee farming was done in the shade of trees that provided a habitat for many animals and insects. Remnant forest trees were used for this purpose, but many species have been planted as well. These include leguminous trees of the genera *Acacia*, *Albizia*, *Cassia*, *Erythrina*, *Gliricidia*, *Inga*, and *Leucaena*, as well as the nitrogen-fixing non-legume sheoaks of the genus *Casuarina*, and the silky oak *Grevillea robusta*.

This method commonly referred to as the traditional shaded method, or "shade-grown". Starting in the 1970s, many farmers switched their production method to sun cultivation, in which coffee is grown in rows under full sun with little or no forest canopy. This causes berries to ripen more rapidly and bushes to produce higher yields, but requires the clearing of trees and increased use of fertilizer and pesticides, which damage the environment and cause health problems.

Unshaded coffee plants grown with fertilizer yield the most coffee, although unfertilized shaded crops generally yield more than unfertilized unshaded crops: the response to fertilizer is much greater in full sun. Although traditional coffee production causes berries to ripen more slowly and produce lower yields, the quality of the coffee is allegedly superior. In addition, the traditional shaded method provides living space for many wildlife species. Proponents of shade cultivation say environmental problems such as deforestation, pesticide pollution, habitat destruction, and soil and water degradation are the side effects of the practices employed in sun cultivation.

The American Birding Association, Smithsonian Migratory Bird Centre, National Arbor Day Foundation, and the Rainforest Alliance have led a campaign for 'shade-grown' and organic coffees, which can be sustainably harvested. Shaded coffee cultivation systems show greater biodiversity than full-sun systems, and those more distant from continuous forest compare rather poorly to undisturbed native forest in terms of habitat value for some bird species.

Another issue concerning coffee is its use of water. According to *New Scientist*, using industrial farming practices, it takes about 140 liters (37 US gal) of water to grow the coffee seeds needed to produce one cup of coffee, and the coffee is often grown in countries where there is a water shortage, such as Ethiopia.

By using sustainable agriculture methods, the amount of water usage can be dramatically reduced, while retaining comparable yields. For comparison, the United States Geological Survey reports that one egg requires an input of 454 liters (120 US gal) of water; one serving of milk requires an input of 246 liters (65 US gal) of water; one serving of rice requires an input of 132 liters (35 US gal) of water; and one glass of wine requires an input of 120 liters (32 US gal) of water.

Coffee grounds may be used for composting or as a mulch. They are especially appreciated by worms and acid-loving plants such as blueberries. Some commercial coffee shops run initiatives to make better use of these grounds, including Starbucks' "Grounds for your Garden" project, and community sponsored initiatives such as "Ground to Ground".

Starbucks sustainability chief Jim Hanna has warned that Climate change may significantly impact coffee yields within a few decades.

Production

Coffee production is the industrial process of converting the raw fruit of the coffee plant into the finished coffee. The cherry has the fruit or pulp removed leaving the seed or bean which is then dried. While all green coffee is processed, the method that is used varies and can have a significant effect on the flavour of roasted and brewed coffee. Coffee production is a major source of income, especially for developing countries where coffee is grown. By adding value, processing the coffee locally, coffee farmers and countries can increase the revenue from coffee.

Picking

A coffee plant usually starts to produce flowers 3–4 years after it is planted, and it is from these flowers that the fruits of the plant (commonly known as coffee cherries) appear, with the first useful harvest possible around 5 years after planting. The cherries ripen around eight months after the emergence of the flower, by changing colour from green to red, and it is at this time that they should be harvested. In most coffee-growing countries, there is one major harvest a year; though in countries like Colombia, where there are two flowerings a year, there is a main and secondary crop, the main one April to June and a smaller one in November to December.

In most countries, the coffee crop is picked by hand, a labour-intensive and difficult process, though in places like Brazil, where the landscape is relatively flat and the coffee fields immense, the process has been mechanized. Whether picked by hand or by machine, all coffee is harvested in one of two ways:

Strip Picked: The entire crop is harvested at one time. This can either be done by machine or by hand. In either case, all of the cherries are stripped off the branch at one time.

Selectively Picked: Only the ripe cherries are harvested and they are picked individually by hand. Pickers rotate among the trees every 8–10 days, choosing only the cherries which are at the peak of ripeness. It usually takes 2-4 years after planting for a coffee plant to produce coffee beans that are ripe enough to harvest. The plant eventually grows small white blossoms that drop and are replaced by green berries. These green berries will become a deep red colour as they get more and more ripe. It takes about 9 months for the green coffee plants to reach their deepest red colour. Because this kind of harvest is labour intensive, and thus more costly, it is used primarily to harvest the finer arabica beans.

The labourers who pick coffee by hand receive payment by the basketful. As of 2003, payment per basket is between US$2.00 to $10 with the overwhelming majority of the labourers receiving payment at the lower end. An experienced coffee picker can collect up to 6-7 baskets a day. Depending on the grower, coffee pickers are sometimes specifically instructed to not pick green coffee berries since the seeds in the berries are not fully formed or mature. This discernment typically only occurs with growers who harvest for higher end/speciality coffee where the pickers are paid better for their labour. Mixes of green and red berries, or just green berries, are used to produce cheaper mass consumer coffee beans, which are characterized by a displeasingly bitter/astringent flavour and a sharp odor. Red berries, with their higher aromatic oil and lower organic acid content, are more fragrant, smooth, and mellow. As such, coffee picking is one of the most important stages in coffee production.

Processing

Wet process:

Figure: *Coffee processing aquapulp*

In the wet process, the fruit covering the seeds/beans is removed before they are dried. Coffee processed by the wet method is called wet processed or washed coffee. The wet method requires the use of specific equipment and substantial quantities of water.

The coffee cherries are sorted by immersion in water.Bad or unripe fruit will float and the good ripe fruit will sink. The skin of the cherry and some of the pulp is removed by pressing the fruit by machine in water through a screen. The bean will still have a significant amount of the pulp clinging to it that needs to be removed. This is done either by the classic ferment-and-wash method or a newer procedure variously called machine-assisted wet processing, aquapulping or mechanical demucilaging:

Figure: *Sorting coffee in water*

In the ferment-and-wash method of wet processing, the remainder of the pulp is removed by breaking down the cellulose by fermenting the beans with microbes and then washing them with large amounts of water. Fermentation can be done with extra water or, in "Dry Fermentation", in the fruit's own juices only.

The fermentation process has to be carefully monitored to ensure that the coffee doesn't acquire undesirable, sour flavours. For most coffees, mucilage removal through fermentation takes between 24 and 36 hours, depending on the temperature, thickness of the mucilage layer and concentration of the enzymes. The end of the fermentation is assessed by feel, as the parchment surrounding the beans loses its slimy texture and acquires a rougher "pebbly" feel. When the fermentation is complete, the coffee is thoroughly washed with clean water in tanks or in special washing machines.

In machine-assisted wet processing, fermentation is not used to separate the bean from the remainder of the pulp; rather, this is done through mechanical scrubbing. This process can cut down on water use and pollution since ferment and wash water stinks. In addition, removing mucilage by machine is easier and more predictable than removing it by fermenting and washing. However, by eliminating the fermentation step and prematurely separating fruit and bean, mechanical demucilaging can remove an important tool that mill operators have of influencing coffee flavour. Furthermore, the ecological criticism of the ferment-and-wash method increasingly has become moot, since a combination of low-water equipment plus settling tanks allows conscientious mill operators to carry out fermentation with limited pollution.

Any wet processing of coffee produces coffee wastewater which can be a pollutant. Ecologically sensitive farms reprocess the wastewater along with the shell and mucilage as compost to be used in soil fertilization programmes. The amount of water used in processing can vary, but most often is used in a 1 to 1 ratio.

After the pulp has been removed what is left is the bean surrounded by two additional layers, the silver skin and the parchment. The beans must be dried to a water content of about 10% before they are stable. Coffee beans can be dried in the sun or by machine but in most cases it is dried in the sun to 12-13% moisture and brought down to 10% by machine. Drying entirely by machine is normally only done where space is at a premium or the humidity is too high for the beans to dry before mildewing.

Figure: *Coffee drying in the sun. Dolka Plantation Costa Rica*

When dried in the sun coffee is most often spread out in rows on large patios where it needs to be raked every six hours to promote even drying and prevent the growth of mildew. Some coffee is dried on large raised tables where the coffee is turned by hand. Drying coffee this way has the advantage of allowing air to circulate better around the beans promoting more even drying but increases cost and labour significantly.

After the drying process (in the sun and/or through machines), the parchment skin or pergamino is thoroughly dry and crumbly, and easily removed in the Hulling process. Coffee occasionally is sold and shipped in parchment or en pergamino, but most often a machine called a huller is used to crunch off the parchment skin before the beans are shipped.

Dry Process

Dry process, also known as unwashed or natural coffee, is the oldest method of processing coffee. The entire cherry after harvest is first cleaned and then placed in the sun to dry on tables or in thin layers on patios:

The harvested cherries are usually sorted and cleaned, to separate the unripe, overripe and damaged cherries and to remove dirt, soil, twigs and leaves. This can be done by winnowing, which is commonly done by hand, using a large sieve. Any unwanted cherries or other material not winnowed away can be picked out from the top of the sieve. The ripe cherries can also be separated by flotation in washing channels close to the drying areas.

The coffee cherries are spread out in the sun, either on large concrete or brick patios or on matting raised to waist height on trestles. As the cherries dry, they are raked or turned by hand to ensure even drying and prevent mildew. It may take up to 4 weeks before the cherries are dried to the optimum moisture content, depending on the weather conditions. On larger plantations, machine-drying is sometimes used to speed up the process after the coffee has been pre-dried in the sun for a few days.

The drying operation is the most important stage of the process, since it affects the final quality of the green coffee. A coffee that has been overdried will become brittle and produce too many broken beans during hulling (broken beans are considered defective beans). Coffee that has not been dried sufficiently will be too moist and prone to rapid deterioration caused by the attack of fungi and bacteria.

The dried cherries are stored in bulk in special silos until they are sent to the mill where hulling, sorting, grading and bagging take place. All the outer layers of the dried cherry are removed in one step by the hulling machine.

The dry method is used for about 90% of the Arabica coffee produced in Brazil, most of the coffees produced in Ethiopia, Haiti and Paraguay, as well as for some Arabicas produced in India and Ecuador. Almost all Robustas are processed by this method. It is not practical in very rainy regions, where the humidity of the atmosphere is too high or where it rains frequently during harvesting.

Semi dry Process

Semi dry is a hybrid process used in Indonesia and Brazil. In Indonesia, the process is also called "wet hulled", "semi-washed" or "Giling Basah". Literally translated from Indonesian, Giling Basah means "wet grinding". This process is said to reduce acidity and increase body.

Most small-scale farmers in Sumatra, Sulawesi, Flores and Papua use the giling basah process. In this process, farmers remove the outer skin from the cherries mechanically, using locally built pulping machines. The coffee beans, still coated with mucilage, are then stored for up to a day. Following this waiting period, the mucilage is washed off and the parchment coffee is partially dried in the sun before sale at 30% to 35% moisture content.

Milling

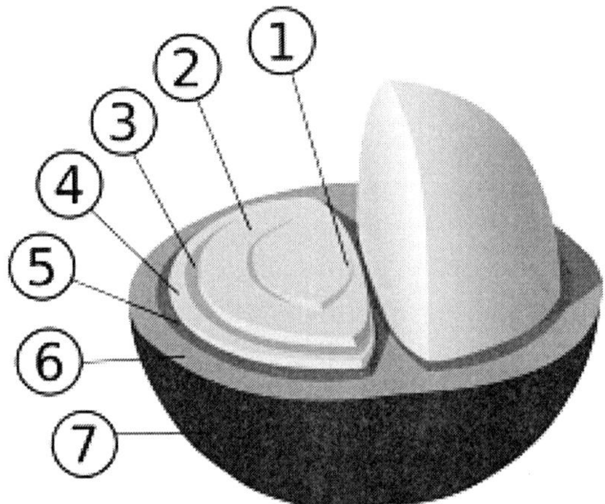

Figure: *Structure of coffee berry and beans: 1: centre cut 2:bean (endosperm) 3: silver skin (testa, epidermis), 4: parchment (hull, endocarp) 5: pectin layer 6: pulp (mesocarp) 7: outer skin (pericarp, exocarp)*

The final steps in coffee processing involve removing the last layers of dry skin and remaining fruit residue from the now dry coffee, and cleaning and sorting it. These steps are often called dry milling to distinguish them from the steps that take place before drying, which collectively are called wet milling.

Hulling

The first step in dry milling is the removal of what is left of the fruit from the bean, whether it is the crumbly parchment skin of wet-processed coffee,the parchment skin and dried mucilage of semi-dry-processed coffee, or the entire dry, leathery fruit covering of the dry-processed coffee. Hulling is done with the help of machines, which can range from simple millstones to sophisticated machines that gently whack at the coffee.

Polishing

This is an optional process in which any silver skin that remains on the beans after hulling is removed in a polishing machine. This is done to improve the appearance of green coffee beans and eliminate a byproduct of roasting called chaff. It is described by some to be detrimental to the taste by raising the temperature of the bean through friction which changes the chemical makeup of the bean.

Cleaning and Sorting

Most fine coffee goes through a battery of machines that sort the coffee by density of bean and by bean size, all the while removing sticks, rocks, nails, and miscellaneous debris that may have become mixed with the coffee during drying. First machines blow the beans into the air; those that fall into bins closest to the air source are heaviest and biggest; the lightest (and likely defective) beans plus chaff are blown in the farthest bin. Other machines shake the beans through a series of sieves, sorting them by size. Finally, a machine called a gravity separator shakes the sized beans on a tilted table, so that the heaviest, densest and best vibrate to one side of the pulsating table, and the lightest to the other.

The final step in the cleaning and sorting procedure is called colour sorting, or separating defective beans from sound beans on the basis of colour rather than density or size. Colour sorting is the trickiest and perhaps most important of all the steps in sorting and cleaning. With most high-quality coffees colour sorting is done in the simplest possible way: by hand. Teams of workers pick discoloured and other defective beans from the sound beans. The very best coffees may be hand-cleaned twice (double picked) or even three times (triple picked). Coffee that has been cleaned by hand is usually called European preparation; most speciality coffees have been cleaned and sorted in this way.

Colour sorting can also be done by machines. Streams of beans fall rapidly, one at a time, past sensors that are set according to parameters that identify defective beans by value (dark to light) or by colour. A tiny, decisive puff of compressed air pops each defective bean out of the stream of sound beans the instant the machine detects an anomaly. However, these machines are currently not used widely in the coffee industry for two reasons. First, the capital investment to install these delicate machines and the technical support to maintain them is daunting. Second, sorting coffee by hand supplies much-needed work for the small rural communities that often cluster around coffee mills. Nevertheless, computerized colour sorters are essential to coffee industries in regions with relatively high standards of living and high wage demands.

Grading

Grading is the process of categorizing coffee beans on the basis of various criteria such as size of the bean, where and at what altitude it was grown, how it was prepared and picked, and how good it tastes, or its cup quality. Coffees also may be graded by the number of imperfections (defective and broken beans, pebbles, sticks, etc.) per sample. For the finest coffees, origin of the beans (farm or estate, region, cooperative) is especially important. Growers of premium estate or cooperative coffees may impose a level of quality control that goes well beyond conventionally defined grading criteria, because they want their coffee to command the higher price that goes with recognition and consistent quality.

Other steps

Aging: All coffee, when it was introduced in Europe, came from the port of Mocha in what is now modern day Yemen. To import the beans to Europe the coffee was on boats for a long sea voyage around the Horn of Africa. This long journey and the exposure to the sea air changed the coffee's flavour. Later, coffee spread to India and Indonesia but still required a long sea voyage. Once the Suez Canal was opened the travel time to Europe was greatly reduced and coffee whose flavour had not changed due to a long sea voyage began arriving. To some degree, this fresher coffee was rejected because Europeans had developed a taste for the changes that were brought on by the long sea voyage. To meet this desire, some coffee was aged in large open-sided warehouses at port for six or more months in an attempt to simulate the effects of a long sea voyage before it was shipped to Europe.

Although it is still widely debated, certain types of green coffee are believed to improve with age; especially those that are valued for their low acidity, such as coffees from Indonesia or India. Several of these coffee producers sell coffee beans that have been aged for as long as 3 years, with some as long as 8 years. However, most coffee experts agree that a green coffee peaks in flavour and freshness within one year of harvest, because over-aged coffee beans will lose much of their essential oil content.

Decaffeination

Decaffeination is the process of extracting caffeine from green coffee beans prior to roasting. The most common decaffeination process used in the United States is supercritical carbon dioxide (CO_2) extraction. In this process, moistened green coffee beans are contacted with large quantities of supercritical CO_2 (CO_2 maintained at a pressure of about 4,000 pounds force per square inch (28 MPa) and temperatures between 90 and 100 °C (194 and 212 °F)), which removes about 97% of the caffeine from the beans. The caffeine is then recovered from the CO_2, typically using an activated carbon adsorption system.

Another commonly used method is solvent extraction, typically using oil (extracted from roasted coffee) or ethyl acetate as a solvent. In this process, solvent is added to moistened green coffee beans to extract most of the caffeine from the beans. After the beans are removed from the solvent, they are steam-stripped to remove any residual solvent. The caffeine is then recovered from the solvent, and the solvent is re-used. The Swiss Water Process is also used for decaffeination. Decaffeinated coffee beans have a residual caffeine content of about 0.1% on a dry basis. Not all facilities have decaffeination operations, and decaffeinated green coffee beans are purchased by many facilities that produce decaffeinated coffee.

Storage

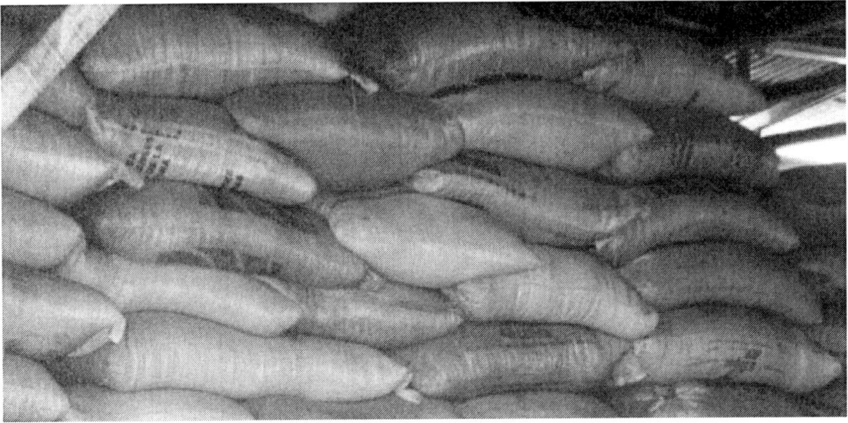

Figure: *Green coffee stored in bags*

Green coffee is usually transported in jute bags. While green coffee may be usable for several years, it is vulnerable to quality degradation based on how it is stored. Jute bags are extremely porous, exposing the coffee to whatever elements it is surrounded by. Coffee that is poorly stored may develop a burlap-like taste known as "bagginess", and its positive qualities may fade.

In recent years, the speciality coffee market has begun to utilize enhanced storage method. A gas barrier liner to jute bags, is sometimes used to preserve the quality of green coffee. Less frequently, green coffee is stored in vacuum packaging; while vacuum packs further reduce the ability of green coffee to interact with oxygen at atmospheric moisture, it is a significantly more expensive storage option.

Roasting

Roasting coffee transforms the chemical and physical properties of green coffee beans into roasted coffee products. The roasting process is what produces the characteristic flavour of coffee by causing the green coffee beans to expand and to change in colour, taste, smell, and density. Unroasted beans contain similar acids, protein, and caffeine as those that have been roasted, but lack the taste. Heat must be applied for the Maillard and other chemical reactions to occur.

As green coffee is more stable than roasted, the roasting process tends to take place close to where it will be consumed. This reduces the time that roasted coffee spends in distribution, giving it a longer shelf life. The vast majority of coffee is roasted commercially on a large scale, but some coffee drinkers roast coffee at home in order to have more control over the freshness and flavour profile of the beans.

Figure: *Light roasted coffee beans*

Process

The coffee roasting process follows coffee processing and precedes coffee brewing. It consists essentially of sorting, roasting, cooling, and packaging but can also include grinding in larger scale roasting houses. In larger operations, bags of green coffee beans are hand or machine-opened, dumped into a hopper, and screened to remove debris. The green beans are then weighed and transferred by belt or pneumatic conveyor to storage hoppers. From the storage hoppers, the green beans are conveyed to the roaster. Roasters typically operate at temperatures between 240–275 °C (464–527 °F), and the beans are roasted for a period of time ranging from 3 to 30 minutes. Initially, the process is endothermic (absorbing heat), but at around 175 °C (347 °F) it becomes exothermic (giving off heat). For the roaster, this means that the beans are heating themselves and an adjustment of the roaster's heat source might be required. At the end of the roasting cycle, the roasted beans are dumped from the roasting chamber and cooled with forced air. Sometimes, in large commercial roasters, the beans are first quenched with a fine water mist. Torrefacto is a roasting process used in Spain and parts of Latin America involving the addition of sugar.

Equipment

The most common roasting machines are of two basic types: drum and hot-air, although there are others including packed bed, tangential and centrifugal roasters. Roasters can operate in either batch or continuous modes. Home roasters are available but less common, and tend to be expensive and time consuming.

Drum machines consist of horizontal rotating drums that tumble the green coffee beans in a heated environment. The heat source can be supplied by natural gas, liquefied petroleum gas (LPG), electricity, or even wood. The most common employ indirectly heated drums where the heat source is under the drum. Direct-fired roasters are roasters in which a flame contacts the beans inside the drum; very few of these machines are still in operation.

Hot-air roasters force heated air through a screen or perforated plate under the coffee beans with sufficient force to lift the beans. Heat is transferred to the beans as they tumble and circulate within this fluidized bed.

Degree of Roasting

Coffee roasters use names for the various degrees of roast, such as City Roast and French Roast, for the internal bean temperatures found during roasting. Roastmasters often prefer to follow a "recipe" or "roast profile" to highlight certain flavour characteristics. Any number of factors may help a person determine the best profile to use, such as the coffee's origin, variety, processing method, or desired flavour characteristics. A roast profile can be presented as a graph showing time on one axis and temperature on the other, which can be recorded manually or using computer software and data loggers linked to temperature probes inside various parts of the roaster.

Determining Degree of Roast

The most popular, but probably the least accurate, method of determining the degree of roast is to judge the bean's colour by eye (the exception to this is using a colourimeter to measure the ground coffee reflectance under infrared light and comparing it to standards such as the Agtron scale). As the beans absorb heat, the colour shifts to yellow and then to increasingly darker shades of brown. During the later stages of roasting, oils appear on the surface of the bean. The roast will continue to darken until it is removed from the heat source. Beans will also darken as they age, making colour alone a poor roast determinant. Most roasters use a combination of bean mass temperature, smell, colour, and sound to monitor the roasting process.

Sound is a good indicator of bean temperature during roasting. There are two temperature thresholds called "cracks" that roasters listen for. At about 205–207 °C (401–405 °F), beans will emit a cracking sound much like popcorn does when it pops, only much quieter. This point is called "first crack," marking the beginning of light roasts. When the beans are at about 224–227 °C (435–441 °F), or a medium roast, they emit a "second crack." This is the dividing point between medium and dark roasts. During first and second "crack" pressure inside the bean has increased to the point where the structure of the bean fractures, rapidly releasing gases, thus an audible sound is emitted.

Degree of Roast Pictorial

These images depict samples taken from the same batch of a typical Brazilian green coffee at various bean temperatures with their subjective roast names and descriptions.

Caffeine Content by Roast Level

Caffeine content varies by roast level. Caffeine diminishes with increased roasting level: light roast, 1.37%; medium roast, 1.31%; and dark roast, 1.31%. However, this does not remain constant in coffee brewed from different grinds and brewing methods. Because the density of coffee changes as it is roasted, different roast levels will contain respectively different caffeine levels when measured by volume or mass, though the bean will still have the same caffeine.

Roast Flavours

At lighter roasts, the bean will exhibit more of its "origin flavour"; the flavours created in the bean by its variety, the soil, altitude, and weather conditions in the location where it was grown.

Coffee beans from famous regions like Java, Kenya, Hawaiian Kona, and Jamaican Blue Mountain are usually roasted lightly so their signature characteristics dominate the flavour. As the beans darken to a deep brown, the origin flavours of the bean are eclipsed by the flavours created by the roasting process itself. At darker roasts, the "roast flavour" is so dominant that it can be difficult to distinguish the origin of the beans used in the roast.

Below, roast levels and their respective flavours are described. These are qualitative descriptions, and thus subjective. As a rule of thumb, the "shinier" the bean is, the more dominant the roasting flavours are.

	Roast level	*Notes*	*Surface*	*Flavor*
Light	*Cinnamon roast, half city, New England*	*After several minutes the beans "pop" or "crack" and visibly expand in size. This stage is called first crack. American mass-market roasters typically stop here.*	*Dry*	*Lighter-bodied, higher acidity, no obvious roast flavor*
Medium	*Full city, American, regular, breakfast, brown*	*After a few short minutes the beans reach this roast, which U.S. specialty sellers tend to prefer.*	*Dry*	*Sweeter than light roast; more body exhibiting more balance in acid, aroma, and complexity. Smoother than the traditional American "medium" roast, but may display fewer of the distinctive taste characteristics of the original coffee.*[11]
Full Roast	*High, Viennese, Continental*	*After a few more minutes the beans begin popping again, and oils rise to the surface. This is called second crack.*	*Slightly shiny*	*Somewhat spicy; complexity is traded for heavier body/mouth-feel. Aromas and flavors of roast become clearly evident.*
Double Roast	*French*	*After a few more minutes or so the beans begin to smoke. The bean sugars begin to carbonize.*	*Very oily*	*Smokey-sweet; light bodied, but quite intense. None of the inherent flavors of the bean are recognizable.*

Home Roasting

Home roasting is the process of roasting small batches of green coffee beans for personal consumption. Even after the turn of the 20th century, it was more common for at-home coffee drinkers to roast their coffee in their residence than it was to buy pre-roasted coffee. Later, home roasting faded in popularity with the rise of the commercial coffee roasting companies. In recent years home roasting of coffee has seen a revival. In some cases there is an economic advantage, but primarily it is a means to achieve finer control over the quality and characteristics of the finished product.

Packaging

Extending the shelflife of roasted coffee relies on maintaining an optimum environment to protect it from exposure to heat, oxygen, and light. Roasted coffee has an optimal typical shelf life of 2 weeks, and ground coffee about 15 minutes. Without some sort of preservation method, coffee becomes stale. The first large scale preservation technique was vacuum packing in cans. However, because coffee emits CO_2 after roasting, coffee to be vacuum packed must be allowed to de-gas for several days before it is sealed. To allow more immediate packaging, pressurized canisters or foil-lined bags with pressure-relief valves can be used. Refrigeration and freezing retards the staling process. Roasted whole beans can be considered fresh for up to one month if kept cool. Once coffee is ground it is best used immediately.

Emissions and Control

Particulate matter (PM), volatile organic compounds (VOC), organic acids, and combustion products are the principal emissions from coffee processing. Several operations are sources of PM emissions, including the cleaning and destoning equipment, roaster, cooler, and instant coffee drying equipment. The roaster is the main source of gaseous pollutants, including alcohols, aldehydes, organic acids, and nitrogen and sulphur compounds. Because roasters are typically natural gas-fired, carbon monoxide (CO) and carbon dioxide (CO_2) emissions result from fuel combustion. Coffee beans also emit carbon monoxide after the roasting process. Decaffeination and instant coffee extraction and drying operations may also be sources of small amounts of VOC. Emissions from the grinding and packaging operations typically are not vented to the atmosphere.

Particulate matter emissions from the roasting and cooling operations are typically ducted to cyclones before being emitted to the atmosphere. Gaseous emissions from roasting operations are typically ducted to a thermal oxidiser or thermal catalytic oxidiser following PM removal by a cyclone. Some facilities use the burners that heat the roaster as thermal oxidisers. However, separate thermal oxidisers are more efficient because the desired operating temperature is typically between 650–816 °C (1202–1501 °F), which is 93–260 °C (199–500 °F) more than the maximum temperature of most roasters. Some facilities use thermal catalytic oxidizers, which require lower operating temperatures to achieve control efficiencies that are equivalent to standard thermal oxidisers. Catalysts are also used to improve the control efficiency of systems in which the roaster exhaust is ducted to the burners that heat the roaster. Emissions from spray dryers are typically controlled by a cyclone followed by a wet scrubber.

The next step in the process is the roasting of the green coffee. Coffee is usually sold in a roasted state, and with rare exceptions all coffee is roasted before it is consumed. It can be sold roasted by the supplier, or it can be home roasted. The roasting process influences the taste of the beverage by changing the coffee seed both physically and chemically. The seed decreases in weight as moisture is lost and increases in volume, causing it to become less dense. The density of the seed also influences the strength of the coffee and requirements for packaging.

The actual roasting begins when the temperature inside the seed reaches approximately 200 °C (392 °F), though different varieties of seeds differ in moisture and density and therefore roast at different rates. During roasting, caramelization occurs as intense heat breaks down starches, changing them to simple sugars that begin to brown, which alters the colour of the seed.

Sucrose is rapidly lost during the roasting process and may disappear entirely in darker roasts. During roasting, aromatic oils and acids weaken, changing the flavour; at 205 °C (401 °F), other oils start to develop. One of these oils, caffeol, is created at about 200 °C (392 °F), which is largely responsible for coffee's aroma and flavour.

Grading the Roasted Seeds

Depending on the colour of the roasted seeds as perceived by the human eye, they will be labeled as light, medium light, medium, medium dark, dark, or very dark. A more accurate method of discerning the degree of roast involves measuring the reflected light from roasted seeds illuminated with a light source in the near infrared spectrum. This elaborate light metre uses a process known as spectroscopy to return a number that consistently indicates the roasted coffee's relative degree of roast or flavour development.

Roast Characteristics

The degree of roast has an effect upon coffee flavour and body. Darker roasts are generally bolder because they have less fibre content and a more sugary flavour. Lighter roasts have a more complex and therefore perceived stronger flavour from aromatic oils and acids otherwise destroyed by longer roasting times. Roasting does not alter the amount of caffeine in the bean, but does give less caffeine when the beans are measured by volume because the beans expand during roasting.

A small amount of chaff is produced during roasting from the skin left on the seed after processing. Chaff is usually removed from the seeds by air movement, though a small amount is added to dark roast coffees to soak up oils on the seeds.

Decaffeination

Decaffeination may also be part of the processing that coffee seeds undergo. Seeds are decaffeinated when they are still green. Many methods can remove caffeine from coffee, but all involve either soaking the green seeds in hot water (often called the "Swiss water process") or steaming them, then using a solvent to dissolve caffeine-containing oils. Decaffeination is often done by processing companies, and the extracted caffeine is usually sold to the pharmaceutical industry.

Storage

Once roasted, coffee seeds must be stored properly to preserve the fresh taste of the seed. Ideally, the container must be airtight and kept in a cool, dry and dark place. In order of importance: air, moisture, heat, and light are the environmental factors responsible for deteriorating flavour in coffee seeds. Folded-over bags, a common way consumers often purchase coffee, are generally not ideal for long-term storage because they allow air to enter. A better package contains a one-way valve, which prevents air from entering. In 1931, a method of vacuum packed cans of coffee was introduced, in which the roasted coffee was packed, 99% of the air was removed and the coffee in the can could be stored indefinitely until the can was opened. Today this method is in mass use for coffee in a large part of the world.

Brewing

Coffee seeds must be ground and brewed to create a beverage. The criteria for choosing a method include flavour and economy. Almost all methods of preparing coffee require the seeds to be ground and mixed with hot water long enough to extract the flavour, but without overextraction that draws out bitter compounds. The spent grounds are removed and the liquid is consumed. There are many brewing variations such as the fineness of grind, the ways in which the water extracts the flavour, additional flavourings (sugar, milk, spices), and spent ground separation techniques. The ideal holding temperature is 79 to 85 °C (174 to 185 °F) and the ideal serving temperature is 68 to 79 °C (154 to 174 °F).

The roasted coffee seeds may be ground at a roastery, in a grocery store, or in the home. Most coffee is roasted and ground at a roastery and sold in packaged form, though roasted coffee seeds can be ground at home immediately before consumption. It is also possible, though uncommon, to roast raw seeds at home.

Roasting has a heavy influence on brewing. Lighter roasted coffee tends to be used for filter coffee as the combination of method and roast style results in higher acidity, complexity and a clearer nuances. Darker roasted coffee is used for espresso because the machine naturally extracts more dissolved solids into the cup resulting in lighter coffee being too acidic.

Coffee seeds may be ground in several ways. A burr grinder uses revolving elements to shear the seed; a blade grinder cuts the seeds with blades moving at high speed; and a mortar and pestle crushes the seeds. For most brewing methods, a burr grinder is deemed superior because the grind is more even and the grind size can be adjusted.

The type of grind is often named after the brewing method for which it is generally used. Turkish grind is the finest grind, while coffee percolator or French press are the coarsest grinds. The most common grinds are between the extremes; a medium grind is used in most common home coffee-brewing machines.

Coffee may be brewed by several methods: boiled, steeped, or pressurized.

Brewing coffee by boiling was the earliest method, and Turkish coffee is an example of this method. It is prepared by grinding or pounding the seeds to a fine powder, then adding it to water and bringing it to the boil for no more than an instant in a pot called a *cezve* or, in Greek, a *bríki*. This produces a strong coffee with a layer of foam on the surface and sediment (which is not meant for drinking) settling on the bottom of the cup.

Coffee percolators and automatic coffeemakers brew coffee using gravity. In an automatic coffeemaker hot water drips onto coffee grounds held in a coffee filter made of paper, plastic, or perforated metal, allowing the water to seep through the ground coffee while extracting its oils and essences. The liquid drips through the coffee and the filter into a carafe or pot, and the spent grounds are retained in the filter.

In a percolator, boiling water is forced into a chamber above a filter by steam pressure created by boiling. The water then seeps through the grounds, and the process is repeated until terminated by removing from the heat, by an internal timer, or by a thermostat that turns off the heater when the entire pot reaches a certain temperature.

Coffee may be brewed by steeping in a device such as a French press (also known as a *cafetière*, coffee press or coffee plunger). Ground coffee and hot water are combined in a cylindrical vessel and left to brew for a few minutes. A circular filter which fits tightly in the cylinder fixed to a plunger is then pushed down from the top to force the grounds to the bottom. Because the coffee grounds are in direct contact with the water, all the coffee oils remain in the beverage, making it stronger and leaving more sediment than in coffee made by an automatic coffee machine. The coffee is poured from the container; the filter retains the grounds at the bottom. 95% of the caffeine is released from the coffee seeds within the first minute of brewing.

The espresso method forces hot pressurized and vaporized water through ground coffee. As a result of brewing under high pressure (ideally between 9–10 atm), the espresso beverage is more concentrated (as much as 10 to 15 times the quantity of coffee to water as gravity-brewing methods can produce) and has a more complex physical and chemical constitution. A well-prepared espresso has a reddish-brown foam called *crema* that floats on the surface. Other pressurized water methods include the moka pot and vacuum coffee maker.

Cold brew coffee is made by steeping coarsely ground seeds in cold water for several hours, then filtering them. This results in a brew lower in acidity than most hot-brewing methods.

Serving

Figure: *Presentation can be an integral part of coffeehouse service, as illustrated by the rosetta design layered into this latte.*

Once brewed, coffee may be served in a variety of ways. Drip-brewed, percolated, or French-pressed/cafetière coffee may be served as *white coffee* with a dairy product such as milk or cream, or dairy substitute, or as *black coffee* with no such addition. It may be sweetened with sugar or artificial sweetener. When served cold, it is called *iced coffee.*

Espresso-based coffee has a wide variety of possible presentations. In its most basic form, an espresso is served simply alone as a *shot*, or *short black*, or with hot water added, known as Caffè Americano. Reversely, long black is made by pouring a double espresso into an equal portion of water, which retains the crema compared to Caffè Americano. Milk is added in various forms to an espresso: steamed milk makes a caffè latte, equal parts steamed milk and milk froth make a cappuccino, and a dollop of hot foamed milk on top creates a caffè macchiato. The use of steamed milk to form patterns such as hearts or maple leaves is referred to as latte art.

Coffee can also be incorporated with alcohol in beverages—it is combined with whiskey in Irish coffee, and forms the base of alcoholic coffee liqueurs such as Kahlúa, and Tia Maria. Coffee is also sometimes used in the brewing process of darker beers, such as a stout or porter.

Instant Coffee

A number of products are sold for the convenience of consumers who do not want to prepare their own coffee.

Instant coffee is dried into soluble powder or freeze-dried into granules that can be quickly dissolved in hot water. Originally invented in 1907, it rapidly gained in popularity in many countries in the post-war period, with Nescafé being the most popular product. Many consumers determined that the convenience in preparing a cup of instant coffee more than made up for a perceived inferior taste. Paralleling (and complementing) the rapid rise of instant coffee was the coffee vending machine, invented in 1947 and multiplying rapidly through the 1950s.

Canned coffee has been popular in Asian countries for many years, particularly in China, Japan, South Korea, and Taiwan. Vending machines typically sell varieties of flavoured canned coffee, much like brewed or percolated coffee, available both hot and cold. Japanese convenience stores and groceries also have a wide availability of bottled coffee drinks, which are typically lightly sweetened and pre-blended with milk. Bottled coffee drinks are also consumed in the United States.

Liquid coffee concentrates are sometimes used in large institutional situations where coffee needs to be produced for thousands of people at the same time. It is described as having a flavour about as good as low-grade robusta coffee, and costs about 10¢ a cup to produce. The machines can process up to 500 cups an hour, or 1,000 if the water is preheated.

Sale and Distribution

***Figure:** Brazilian coffee sacks*

Coffee ingestion on average is about a third of that of tap water in North America and Europe. Worldwide, 6.7 million metric tons of coffee were produced annually in 1998–2000, and the forecast is a rise to seven million metric tons annually by 2010.

Brazil remains the largest coffee exporting nation, but Vietnam tripled its exports between 1995 and 1999 and became a major producer of robusta seeds. Indonesia is the third-largest coffee exporter overall and the largest producer of washed arabica coffee. Organic Honduran coffee is a rapidly growing emerging commodity owing to the Honduran climate and rich soil.

In 2013 The Seattle Times reported that global coffee prices have dropped more than 50 percent year-over-year.

Commodity

Coffee is bought and sold by roasters, investors, and price speculators as a tradable commodity in commodity markets and exchange-traded funds. Coffee futures contracts for Grade 3 washed arabicas are traded on the New York Mercantile Exchange under ticker symbol KC, with contract deliveries occurring every year in March, May, July, September, and December. Coffee is an example of a product that has been susceptible to significant commodity futures price variations.

Higher and lower grade arabica coffees are sold through other channels. Futures contracts for robusta coffee are traded on the London International Financial Futures and Options Exchange and, since 2007, on the New York IntercontinentalExchange. Coffee has been described by many, including historian Mark Pendergrast, as the world's "second most legally traded commodity." However, this claim has been recently refuted by Pendergrast among others after further research.

Health and pharmacology

Method of action:

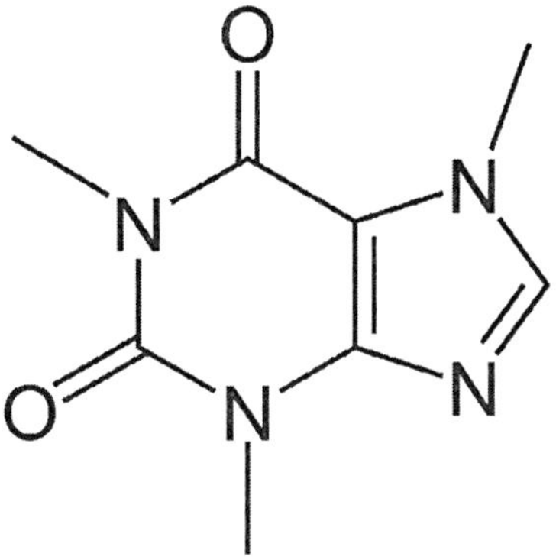

Figure: Skeletal structure of a caffeine molecule

The primary psychoactive chemical in coffee is caffeine, an adenosine antagonist that is known for its stimulant effects. Coffee also contains the monoamine oxidase inhibitors â-carboline and harmane, which may contribute to its psychoactivity.

In a healthy liver, caffeine is mostly broken down by the hepatic microsomal enzymatic system. The excreted metabolites are mostly paraxanthines—theobromine and theophylline—and a small amount of unchanged caffeine. Therefore, the metabolism of caffeine depends on the state of this enzymatic system of the liver.

General Health

Extensive scientific research has been conducted to examine the relationship between coffee consumption and an array of medical conditions. The general consensus in the medical community is that moderate regular coffee drinking in healthy individuals is either essentially benign or mildly beneficial. In 2012, the National Institutes of Health–AARP Diet and Health Study analysed the relationship between coffee drinking and mortality. They found that the amount of coffee consumed correlated negatively with risk of death, and that those who drank any coffee lived longer than those who did not. However the authors noted, "whether this was a causal or associational finding cannot be determined from our data." A similar study with similar results was published in the New England Journal of Medicine in 2012. Researchers involved in an ongoing 22-year study by the Harvard School of Public Health stated that "the overall balance of risks and benefits [of coffee consumption] are on the side of benefits."

Findings have also been contradictory as to whether coffee has any specific health benefits, and results are similarly conflicting regarding the potentially harmful effects of coffee consumption. Furthermore, results and generalizations are complicated by differences in age, gender, health status, and serving size.

Health Benefits

According to Cancer Research UK, the results of the large-scale study published in 2012 clarified the evidence of the effect of coffee drinking on cancer, and that evidence beforehand had been conflicting. They say the study showed that drinking coffee "had no effect on the risk of dying from cancer."

Other studies suggest coffee consumption reduces the risk of Alzheimer's disease, dementia, Parkinson's disease, heart disease, diabetes mellitus type 2, non-alcoholic fatty liver disease, cirrhosis, and gout.

The fact that decaffeinated coffee also exhibits preventative effects against diseases such as prostate cancer and type 2 diabetes suggests that coffee's health benefits are not solely a product of its caffeine content. Specifically, the antidiabetic effect of caffeine has been attributed to caffeic acid and chlorogenic acid.

The presence of antioxidants in coffee has been shown to prevent free radicals from causing cell damage. Evidence suggests that roasted coffee has a stronger antioxidant effect than green coffee.

Coffee is no longer thought to be a risk factor for coronary heart disease. A 2012 meta-analysis concluded that people who drank moderate amounts of coffee had a lower rate of heart failure, with the biggest effect found for those who drank more than four cups a day.

Caffeine acts as an acute antidepressant. A review published in 2004 indicated a negative correlation between suicide rates and coffee consumption. It was suggested that the action of caffeine in blocking the inhibitory effects of adenosine on dopamine nerves in the brain reduced feelings of depression. Coffee consumption is also associated with improved endothelial function. Coffee extracts have been shown to inhibit 11beta-hydroxysteroid dehydrogenase type 1, enzyme which converts cortisone to cortisol and is a current pharmaceutical target for the treatment of diabetes type 2 and the metabolic syndrome.

Social and Culture

Coffee is often consumed alongside (or instead of) breakfast by many at home. It is often served at the end of a meal, normally with a dessert, and at times with an after-dinner mint especially when consumed at a restaurant or dinner party.

Aggressively promoted by the Pan-American Coffee Bureau, the "coffee break" was first promoted in 1952. Hitherto unknown in the workplace, its uptake was facilitated by the recent popularity of both instant coffee and vending machines, and has become an institution of the American workplace.

Coffeehouses

Most widely known as coffeehouses or cafés, establishments serving prepared coffee or other hot beverages have existed for over five hundred years.

Various legends involving the introduction of coffee to Istanbul at a "Kiva Han" in the late 15th century circulate in culinary tradition, but with no documentation.

Coffeehouses in Mecca soon became a concern as places for political gatherings to the imams who banned them, and the drink, for Muslims between 1512 and 1524. In 1530 the first coffee house was opened in Damascus. First coffee houses in Constantinople was opened in 1475 by traders arriving from Damascus and Aleppo. Soon after, coffee houses became part of the Ottoman Culture, spreading rapidly to all regions of the Ottoman Empire.

In the 17th century, coffee appeared for the first time in Europe outside the Ottoman Empire, and coffeehouses were established and quickly became popular. The first coffeehouses in Western Europe appeared in Venice, a result of the traffic between La Serenissima and the Ottomans; the very first one is recorded in 1645. The first coffeehouse in England was set up in Oxford in 1650 by a Jewish man named Jacob in the building now known as "The Grand Cafe". A plaque on the wall still commemorates this and the Cafe is now a trendy cocktail bar. By 1675, there were more than 3,000 coffeehouses in England.

After the second Turkish siege of Vienna in 1683, the Viennese discovered many bags of coffee in the abandoned Ottoman encampment.

Using this captured stock, a Polish soldier named Kulczycki opened the first coffeehouse in Vienna.

In 1672 an Armenian named Pascal established a coffee stall in Paris that was ultimately unsuccessful and the city had to wait until 1689 for its first coffeehouse when Procopio Cutò opened the Café Procope. This coffeehouse still exists today and was a major meeting place of the French Enlightenment; Voltaire, Rousseau, and Denis Diderot frequented it, and it is arguably the birthplace of the *Encyclopédie*, the first modern encyclopedia. America had its first coffeehouse in Boston, in 1676. Coffee, tea and beer were often served together in establishments which functioned both as coffeehouses and taverns; one such was the Green Dragon in Boston, where John Adams, James Otis and Paul Revere planned rebellion.

The modern espresso machine was born in Milan in 1945 by Achille Gaggia, and from there spread across coffeehouses and restaurants across Italy and the rest of Europe and North America in the early 1950s. An Italian named Pino Riservato opened the first espresso bar, the Moka Bar, in Soho in 1952, and there were 400 such bars in London alone by 1956. Cappucino was particularly popular among English drinkers. Similarly in the United States, the espresso craze spread. North Beach in San Francisco saw the opening of the Caffe Trieste in 1957, which saw Beat Generation poets such as Allen Ginsberg and Bob Kaufman alongside bemused Italian immigrants. Similar such cafes existed in Greenwich Village and elsewhere.

The first Peet's Coffee & Tea store opened in 1966 in Berkeley, California by Dutch native Alfred Peet. He chose to focus on roasting batches with fresher, higher quality seeds than was the norm at the time. He was a trainer and supplier to the founders of Starbuck's.

The international coffeehouse chain Starbucks began as a modest business roasting and selling quality coffee seeds in 1971, by three college students Jerry Baldwin, Gordon Bowker and Zev Siegl. The first store opened on March 30, 1971 at the Pike Place Market in Seattle, followed by a second and third over the next two years. Entrepreneur Howard Schultz joined the company in 1982 as Director of Retail Operations and Marketing, and pushed to sell premade espresso coffee. The others were reluctant, but Schultz opened Il Giornale in Seattle in April 1986. He bought the other owners out in March 1987 and pushed on with plans to expand—from 1987 to the end of 1991, the chain (rebranded from Il Giornale to Starbucks) expanded to over 100 outlets. The company has 16,600 stores in over 40 countries worldwide.

South Korea experienced almost 900 percent growth in the number of coffee shops in the country between 2006 and 2011. The capital city Seoul now has the highest concentration of coffee shops in the world, with more than 10,000 cafes and coffee houses.

Freezing of Fruits and Vegetables

Freezing is one of the oldest and most widely used methods of food preservation, which allows preservation of taste, texture, and nutritional value in foods better than any other method. The freezing process is a combination of the beneficial effects of low temperatures at which microorganisms cannot grow, chemical reactions are reduced, and cellular metabolic reactions are delayed (Delgado and Sun, 2000).

The Importance of Freezing as a Preservation Method

Freezing preservation retains the quality of agricultural products over long storage periods. As a method of long-term preservation for fruits and vegetables, freezing is generally regarded as superior to canning and dehydration, with respect to retention in sensory attributes and nutritive properties (Fennema, 1977). The safety and nutrition quality of frozen products are emphasized when high quality raw materials are used, good manufacturing practices are employed in the preservation process, and the products are kept in accordance with specified temperatures.

The Need for Freezing and Frozen Storage

Freezing has been successfully employed for the long-term preservation of many foods, providing a significantly extended shelf life. The process involves lowering the product temperature generally to -18 °C or below (Fennema *et al.*, 1973). The physical state of food material is changed when energy is removed by cooling below freezing temperature. The extreme cold simply retards the growth of microorganisms and slows down the chemical changes that affect quality or cause food to spoil (George, 1993).

Competing with new technologies of minimal processing of foods, industrial freezing is the most satisfactory method for preserving quality during long storage periods (Arthey, 1993). When compared in terms of energy use, cost, and product quality, freezing requires the shortest processing time. Any other conventional method of preservation focused on fruits and vegetables, including dehydration and canning, requires less energy when compared with energy consumption in the freezing process and storage. However, when the overall cost is estimated, freezing costs can be kept as low (or lower) as any other method of food preservation (Harris and Kramer, 1975).

Current Status of Frozen Food Industry in U.S. and Other Countries

The frozen food market is one of the largest and most dynamic sectors of the food industry. In spite of considerable competition between the frozen food industry and other sectors, extensive quantities of frozen foods are being consumed all over the world. The industry has recently grown to a value of over US$ 75 billion in the U.S. and Europe combined. This number has reached US$ 27.3 billion in 2001 for total retail sales of frozen foods in the U.S. alone (AFFI, 2003). In Europe, based on U.S. currency, frozen food consumption also reached 11.1 million tons in 13 countries in the year 2000 (Quick Frozen Foods International, 2000). Table 1 represents the division of frozen food industry in terms of annual sales in 2001.

Advantages of Freezing Technology in Developing Countries

Developed countries, mostly the U.S., dominate the international trade of fruits and vegetables. The U.S. is ranked number one as both importer and exporter, accounting for the highest percent of fresh produce in world trade. However, many developing countries still lead in the export of fresh exotic fruits and vegetables to developed countries.

For developing countries, the application of freezing preservation is favourable with several main considerations. From a technical point of view, the freezing process is one of the most convenient and easiest of food preservation methods, compared with other commercial preservation techniques. The availability of different types of equipment for several different food products results in a flexible process in which degradation of initial food quality is minimal with proper application procedures. As mentioned earlier, the high capital investment of the freezing industry usually plays an important role in terms of economic feasibility of the process in developing countries. As for cost distribution, the freezing process and storage in terms of energy consumption constitute approximately 10 percent of the total cost. Depending on the government regulations, especially in developing countries, energy cost for producers can be subsidized by means of lowering the unit price or reducing the tax percentage in order to enhance production. Therefore, in determining the economical convenience of the process, the cost related to energy consumption (according to energy tariffs) should be considered.

Increasing Consumer Demand in Developing Countries Due to Modernization

The proportion of fresh food preserved by freezing is highly related to the degree of economic development in a society. As countries become wealthier, their demand for high-valued commodities increases, primarily due to the effect of income on the consumption of high-valued commodities in developing countries. The commodities preserved by freezing are usually the most perishable ones, which also have the highest price. Therefore, the demand for these commodities is less in developing areas. Besides, the need for adequate technology for freezing process is the major drawback of developing countries in competing with industrialized countries. The frozen food industry requires accompanying developments and facilities for transporting, storing, and marketing their products from the processing plant to the consumer (Mallett, 1993). Thus, a large amount of capital investment is needed for these types of facilities. For developing countries, especially in rural or semi-rural areas, the frozen food industry has therefore not been developed significantly compared to other countries.

In recent years, due to the changing consumer profile, the frozen food industry has changed significantly. The major trend in consumer behaviour documented over the last half century has been the increase in the number of working women and the decline in the family size. These two factors resulted in a reduction in time spent preparing food. The entry of more women into the workforce also led to improvements in kitchen appliances and increased the variability of ready-to-eat or frozen foods available in the market. Besides, the increased usage of microwave ovens, affecting food habits in general and the frozen food market in particular, as well as allowing rapid preparation of meals and greater flexibility in meal preparation. The frozen food industry is now only limited by imagination, an output of which increases continuously to supply the increasing demand for frozen products and variability.

Market Share of Frozen Fruits and Vegetables

Today in modern society, frozen fruits and vegetables constitute a large and important food group among other frozen food products (Arthey, 1993). The historical development of commercial freezing systems designed for special food commodities helped shape the frozen food market. Technological innovations as early as 1869 led to the commercial development and marketing of some frozen foods. Early products saw limited distribution through retail establishments due to insufficient supply of mechanical refrigeration. Retail distribution of frozen foods gained importance with the development of commercially frozen vegetables in 1929.

The frozen vegetable industry mostly grew after the development of scientific methods for blanching and processing in the 1940s. Only after the achievement of success in stopping enzymatic degradation, did frozen vegetables gain a strong retail and institutional appeal. Today, market studies indicate that considering overall consumption of frozen foods, frozen vegetables constitute a very significant proportion of world frozen-food categories (excluding ice cream) in Austria, Denmark, Finland, France, Germany, Italy, Netherlands, Norway, Sweden, Switzerland, UK, and the USA. The division of frozen vegetables in terms of annual sales in 2001 is shown in Table 3.

Commercialization history of frozen fruits is older than frozen vegetables. The commercial freezing of small fruits and berries began in the eastern part of the U.S. in about 1905. The main advantage of freezing preservation of fruits is the extended usage of frozen fruits during off-season. Additionally, frozen fruits can be transported to remote markets that could not be accessed with fresh fruit. Also, freezing preservation makes year-round further processing of fruit products possible, such as jams, juice, and syrups from frozen whole fruit, slices, or pulps. In summary, the preservation of fruits by freezing has clearly become one the most important preservation methods.

Future Trends in Freezing Technology

The frozen food industry is highly based in modern science and technology. Starting with the first historical development in freezing preservation of foods, today, a combination of several factors influences the commercialization and usage of freezing technology. The future growth of frozen foods will mostly be affected by economical and technological factors. Growth in population, personal incomes, relative cost of other forms of foods, changes in tastes and preferences, and technological advances in freezing methods are some of the factors concerned with the future of freezing technology.

Population growth and increasing demand for food has generated the need for commercial production of food commodities in large-scale operations. Thus, availability of proper equipment suitable for continuous processing would be valuable for freezing preservation methods. In addition depending on personal incomes, relative cost of frozen products is one of the most important of economical factors. Producing the highest quality at the lowest cost possible is highly dependent on the technology used. As a result, developments in freezing technology in recent years have mostly been characterized by the improvements in mechanical handling and process control to increase freezing rate and reduce cost.

Today an increasing demand for frozen foods already exits and further expansion of the industry is primarily dependent on the ability of food processors to develop higher qualities in both process techniques and products. Improvements can only be achieved by focusing on new technologies and investigating poorly understood factors that influence the quality of frozen food products. Improvements in new and convenient forms of foods, as well as more information on relative cost and nutritive values of frozen foods, will contribute toward continued growth of the industry.

Freezing Technology

Freezing has long been used as a method of preservation, and history reveals it was mostly shaped by the technological developments in the process. A small quantity of ice produced without using a "natural cold" in 1755 was regarded as the first milestone in the freezing process. Firstly, ice-salt systems were used to preserve fish and later on, by the late 1800's, freezing was introduced into large-scale operations as a method of commercial preservation. Meat, fish, and butter, the main products preserved in this early example, were frozen in storage chambers and handled as bulk commodities (Persson and Lohndal, 1993).

In the following years, scientists and researchers continuously worked to achieve success with commercial freezing trials on several food commodities. Among these commodities, fruits were one of the most important since freezing during the peak growing season had the advantage of preserving fruit for later processing into jams, jellies, ice cream, pies, and other bakery foods. Although commercial freezing of small fruits and berries first began around 1905 in the eastern part of the United States, the commercial freezing of vegetables is much more recent. Starting from 1917, only private firms conducted trials on freezing vegetables, but achieving good quality in frozen vegetables was not possible without pre-treatments due to the enzymatic deterioration. In 1929, the necessity of blanching to inactivate enzymes before freezing was concluded by several researchers to avoid deterioration and off-flavours caused by enzymatic degradation.

The modern freezing industry began in 1928 with the development of double-belt contact freezers by a technologist named Clarence Birdseye. After the revolution in the quick freezing process and equipment, the industry became more flexible, especially with the usage of multi-plate freezers. The earlier methods achieved successful freezing of fish and poultry, however with the new quick freezing system, packaged foods could be frozen between two metal belts as they moved through a freezing tunnel. This improvement was a great advantage in the commercial large-scale freezing of fruits and vegetables. Furthermore, quick-freezing of consumer-size packages helped frozen vegetables to be accepted rapidly in late 1930s.

Today, freezing is the only large-scale method that bridges the seasons, as well as variations in supply and demand of raw materials such as meat, fish, butter, fruits, and vegetables. Besides, it makes possible movement of large quantities of food over geographical distances (Persson and Londahl, 1993). It is important to control the freezing process, including the pre-freezing preparation and post-freezing storage of the product, in order to achieve high-quality products (George, 1993). Therefore, the theory of the freezing process and the parameters involved should be understood clearly.

Freezing Process

The freezing process mainly consists of thermodynamic and kinetic factors, which can dominate each other at a particular stage in the freezing process. Major thermal events are accompanied by reduction in heat content of the material during the freezing process. The material to be frozen first cools down to the temperature at which nucleation starts. Before ice can form, a nucleus, or a seed, is required upon which the crystal can grow; the process of producing this seed is defined as nucleation. Once the first crystal appears in the solution, a phase change occurs from liquid to solid with further crystal growth. Therefore, nucleation serves as the initial process of freezing, and can be considered as the critical step that results in a complete phase change.

Freezing Point of Foods

Freezing point is defined as the temperature at which the first ice crystal appears and the liquid at that temperature is in equilibrium with the solid. If the freezing point of pure water is considered, this temperature will correspond to 0 °C (273°K). However, when food systems are frozen, the process becomes more complex due to the existence of both free and bound water. Bound water does not freeze even at very low temperatures. Unfreezable water contains soluble solids, which cause a decrease in the freezing point of water lower than 0 °C. During the freezing process, the concentration of soluble solids increases in the unfrozen water, resulting in a variation in freezing temperature. Therefore, the temperature at which the first ice crystal appears is commonly regarded as the initial freezing temperature. There are empirical equations in literature that can calculate the initial freezing temperature of certain foods as a function of their moisture content.

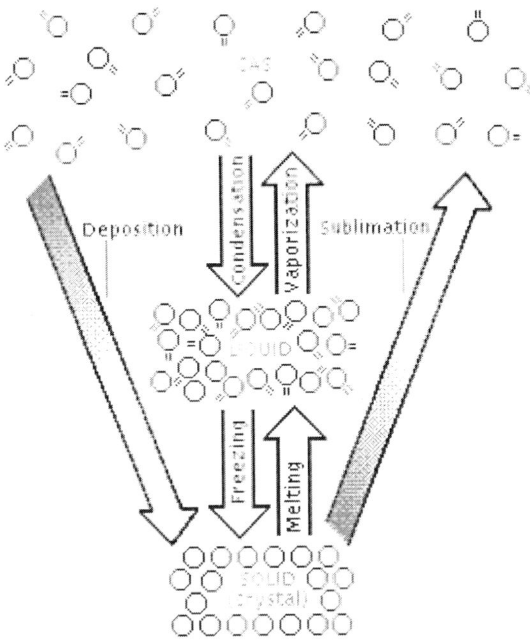

Figure: *A schematic illustration of overall freezing process.*

There are several methods of food freezing, and depending on the method used, the quality of the frozen food may vary. However, regardless of the method chosen, the main principle behind all freezing processes is the same in terms of process parameters. The International Institute of Refrigeration (IIR) has provided definitions to establish a basis for the freezing process. According to their definition, the freezing process is basically divided into three stages based on major temperature changes in a particular location in the product.

Beginning with the prefreezing stage, the food is subjected to the freezing process until the appearance of the first crystal. If the material frozen is pure water, the freezing temperature will be 0 °C and, up to this temperature, there will be a subcooling until the ice formation begins. In the case of foods during this stage, the temperature decreases to below freezing temperature and, with the formation of the first ice crystal, increases to freezing temperature. The second stage is the freezing period; a phase change occurs, transforming water into ice. For pure water, temperature at this stage is constant; however, it decreases slightly in foods, due to the increasing concentration of solutes in the unfrozen water portion. The last stage starts when the product temperature reaches the point where most freezable water has been converted to ice, and ends when the temperature is reduced to storage temperature (Persson and Lohndal, 1993).

The freezing time and freezing rate are the most important parameters in designing freezing systems. The quality of the frozen product is mostly affected by the rate of freezing, while time of freezing is calculated according to the rate of freezing. For industrial applications, they are the most essential parameters in the process when comparing different types of freezing systems and equipment (Persson and Lohndal, 1993).

TEMPERATURE

0°C

PREFREEZING FREEZING REDUCTION TO STORAGE TEMPERATURE TIME

Figure: *Practical definition of the freezing process for pure water (Mallett, 1993).*

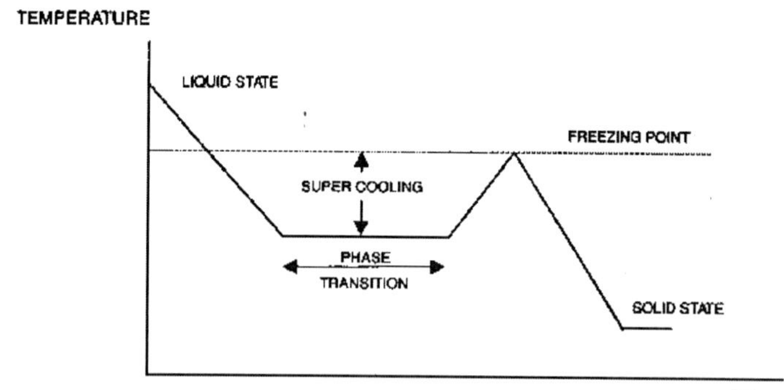

Figure: *Practical definition of the freezing process for foods (Mallett, 1993).*

Refrigeration

Refrigeration is defined as the elimination of heat from a material at a temperature higher than the temperature of its surroundings. The mechanism of refrigeration is a part of the freezing process and freezing storage involved in the thermodynamic aspects of freezing. According to the second law of thermodynamics, heat only flows from higher to lower temperatures. Therefore, in order to raise the heat from a lower to a higher temperature level, expenditure of work is needed. The aim of industrial refrigeration processes is to eliminate heat from low temperature points towards points with higher temperature. For this reason, either closed mechanical refrigeration cycles in which refrigeration fluids circulate, or open cryogenic systems with liquid nitrogen (LIN) or carbon dioxide (CO_2), are commonly used by the food industry.

The main elements in a closed mechanical refrigeration system are the condenser, compressor, evaporator, and the expansion valve. The refrigerants hydrochlorofluorocarbon (HCFC) and ammonia are examples of the refrigerants circulated in these types of mechanical refrigeration systems. A simple scheme for the closed mechanical refrigeration system.

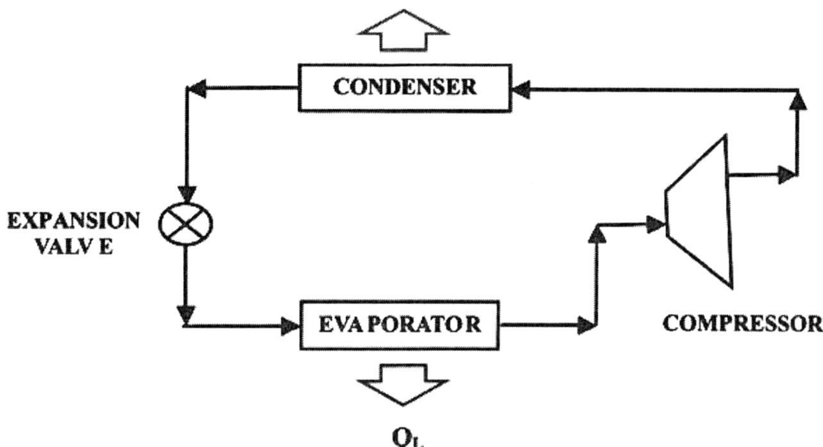

Figure: *A simple scheme for a one-stage closed mechanical refrigeration system. (Adapted from Stoecker, W.F. and Jones J.W., Refrigeration and Air Conditioning, McGraw-Hill, New York, 1982)*

Starting at the suction point of the compressor, fluid in a vapor state is compressed into the compressor where an increase in temperature and pressure takes place. The fluid then flows through the condenser where it decreases in energy by giving off heat and converting to a liquid state. After the phase, a change occurs inside the condenser, the fluid flows through the expansion valve where the pressure decreases to convert liquid into a form of liquid-gas mixture. Finally, the liquid-gas mixture flows through the evaporator where it is converted into a saturated vapor state and removes heat from the environment in the process of cooling. With this last stage the loop restarts again.

The other refrigeration system employed by the food industry is the cryogenic system with carbon dioxide or liquid nitrogen. The refrigerant in this system is consumed differently from the circulating fluid in closed mechanical systems.

Refrigerants

There are several refrigerants available for refrigeration systems. The selection of a proper refrigerant is based on physical, thermodynamic, and chemical properties of the fluid. Environmental considerations are also important in refrigerant selection, since leaks within the system produce deleterious effects on the atmospheric ozone layer. Some refrigerants, including halocarbons, have been banned to avoid potential hazardous effects (Stoecker and Jones, 1982). For industrial applications, ammonia is commonly used, while chlorofluoromethane and tetrafluoroethane are also recommended as refrigerants (Persson and Lohndal, 1993).

Freezing Capacity

Freezing equipment selection is based on the requirements for freezing a certain quantity of food per hour. For any type of freezer, freezing capacity (expressed in tonnes per hour) is defined as the ratio of the quantity of the product that can be loaded into the freezer to the holding time of the product in that particular freezer. The first parameter, the amount of food product loaded into the freezer, is affected by both the dimensions of the product and the mechanical constraints of the freezer. The denominator (holding time) has an important role in freezing systems and is based on the calculation of the amount of heat removed from the product per hour, which varies depending on the type of product frozen (Persson and Lohndal, 1993).

Freezing Systems

There is a variety of freezing systems available for freezing, and for most products, more than one type of freezer can be used. Therefore, in selecting a freezing system initially, a cost-benefit analysis should be conducted based on three important factors: economics, functionality, and feasibility. Financial considerations mainly involve capital investment and the production cost of selected equipment. Product losses during freezing operation should be included in cost estimation since generating higher cost freezers may have other benefits in terms of reducing product losses. Functional factors are mostly based on the suitability of the selected freezer for particular products. The mode of process, either in-line or batch, should be considered based on the fact that computerized systems are becoming more important for ease of handling and lowering production costs. Mechanical constraints for the freezer should also be considered since some types of freezers are not physically suitable for freezing certain products. Lastly, the feasibility of the process should be considered in terms of plant location or location of the processing area, as well as cleanability and hygienic design, and desired product quality (Johnston *et al.,* 1994).

These factors and initial considerations can help eliminate several choices in freezer selection, but the relative importance of factors may change depending on the process. For developing countries where the freezing application is relatively new, the cost factor becomes more important than other factors due to the decreased production rates and need for lower capital investment costs.

Freezing Equipment

The industrial equipment for freezing can be categorized in many ways, namely as equipment used for batch or in-line operation, heat transfer systems (air, contact, cryogenic), and product stability. The rate of heat transfer from the freezing medium to the product is important in defining the freezing time of the product. Therefore, the equipment selected for freezing process characterizes the rate of freezing.

Air-blast Freezers

The air blast freezer is one the oldest and commonly used freezing equipment due to its temperature stability and versatility for several product types. In general, air is used as the freezing medium in the freezing design, either as still air or forced air. Freezing is accomplished by placing the food in freezing rooms called sharp freezers. Still, air freezing is the cheapest way of freezing and has the added advantage of a constant temperature during frozen storage, which allows usage for unprocessed bulk products like beef quarters and fish. However, it is the slowest method of freezing due to the low surface heat transfer coefficient of circulating air inside the room. Freezing time in sharp freezers is largely dependent on the temperature of the freezing chamber and the type, initial temperature, and size of product. An improved version of the still air freezer is the forced air freezer, which consists of air circulation by convection inside the freezing room. However, even modification of the sharp freezer with extra refrigeration capacity and fans for increased air circulation does not help control the air flow over the products during slow freezing.

There are a considerable number of designs and arrangements for air blast freezers, primarily grouped in two categories depending on the mode of process, as either inline or batch. Continuous freezers are the most suitable systems for mass production of packaged products with similar freezing times, in which the product is carried through on trucks or on conveyors. The system works on a semi-batch principle when trucks are used, since they remain stationary during the process except when a new truck enters one end of the tunnel, thus moving the others along to release a finished one at the exit. The batch freezers are more flexible since a variety of products can be frozen at the same time on individual trolleys. Over-loading may be a problem for these types of freezers, thus the process requires closer supervision than continuous systems.

Tunnel Freezers

In tunnel freezers, the products on trays are placed in racks or trolleys and frozen with cold air circulation inside the tunnel. In order to allow air circulation, optimum space is provided between layers of trolley, which can be moved continuously in and out of the freezer manually or by forklift trucks. This freezing system is suitable for all types of products, although there are some mechanical constraints including the requirement of high manpower for handling, cleaning, and transportation of trays.

Belt Freezers

Belt freezers were first designed to provide continuous product flow with the help of a wire mesh conveyor inside the blast rooms. A poor heat transfer mechanism and the mechanical problems were solved in modern belt freezers by providing a vertical airflow to force air through the product layer. Airflow has good contact with the product only when the entire product is evenly distributed over the conveyor belt. In order to decrease required floor space, the belts can be arranged in a multi-tier belt freezer or a spiral belt freezer. Spiral belt freezers consist of a belt that can be bent laterally around a rotating drum to maximize belt surface area in a given floor space. This type of design has the advantage of eliminating product damage in transfer points, especially for products that require gentle handling. Both packed and unpacked products with long freezing times (10 min to 3 hr) can be frozen in spiral belt freezers due to the flexibility of the equipment.

Fluidized bed Freezers

The fluidized bed freezer, a fairly recent modified type of air-blast freezer for particular product types, consists of a bed with a perforated bottom through which cold air is blown vertically upwards. The system relies on forced cold air from beneath the conveyor belt, causing the products to suspend or float in the cold air stream. The use of high air velocity is very effective for freezing unpacked foods, especially when they can be completely surrounded by flowing air, as in the case of fluidized bed freezers.

The use of fluidization has several advantages compared with other methods of freezing since the product is individually quick frozen (IQF), which is convenient for particles with a tendency to stick together (Persson and Lohndal, 1993). The idea of individually quick frozen foods (IQF) started with the first technological developments aimed at quick freezing. The need for an effective means of freezing small particles with the potential for lumping during the process is the objective of IQF freezing. Small vegetables, prawns, shrimp, french-fried potatoes, diced meat, and fruits are some of the products now frozen with this technology.

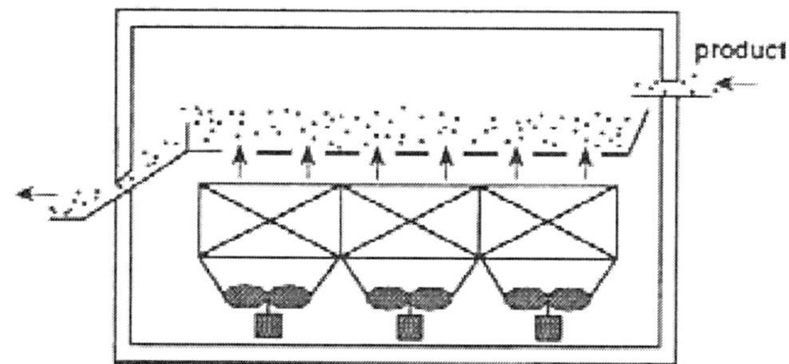

Figure: *Simple working principle of a fluidized bed freezer.*

Contact Freezers

Contact freezing is the one of the most efficient ways of freezing in terms of heat transfer mechanism. In the process of freezing, the product can be in direct or indirect contact with the freezing medium. For direct contact freezers, the product being frozen is fully surrounded by the freezing medium, the refrigerant, maximizing the heat transfer efficiency. A schematic illustration. For indirect contact freezers, the product is indirectly exposed to the freezing medium while in contact with the belt or plate, which is in contact with the freezing medium.

Immersion Freezers

The immersion freezer consists of a tank with a cooled freezing media, such as glycol, glycerol, sodium chloride, calcium chloride, and mixtures of salt and sugar. The product is immersed in this solution or sprayed while being conveyed through the freezer, resulting in fast temperature reduction through direct heat exchange (Hung and Kim, 1996). Direct immersion of a product into a liquid refrigerant is the most rapid way of freezing since liquids have better heat conducting properties than air. The solute used in the freezing system should be safe without taste, odour, colour, or flavour, and for successful freezing, products should be greater in density than the solution. Immersion freezing systems have been commonly used for shell freezing of large particles due to the reducing ability of product dehydration when the outer layer is frozen quickly. A commonly seen problem in these freezing systems is the dilution of solution with the product, which can change the concentration and process parameters. Thus, in order to avoid product contact with the liquid refrigerant, flexible membranes can be used.

Indirect Contact Freezers

In this type of freezer, materials being frozen are separated from the refrigerant by a conducting material, usually a steel plate. The mechanism of indirect contact freezer. Indirect contact freezers generally provide an efficient medium for heat transfer, although the system has some limitations, especially when used for packaged foods due to resistance of package to heat transfer. Additionally, corrosive effects may occur due to interaction of metal packages with heat transfer surfaces.

Plate Freezers

The most common type of contact freezer is the plate freezer. In this case, the product is pressed between hallow metal plates, either horizontally or vertically, with a refrigerant circulating inside the plates. Pressure is applied for good contact as schematically.

This type of freezing system is only limited to regular-shaped materials or blocks like beef patties or block-shaped packaged products.

Contact belt Freezers

This type of freezer is designed with single-band or double-band for freezing of thin product layers. The design can be either straight forward or drum. Typical products frozen in belt freezers are, fruit pulps, egg yolk, sauces and soups.

Cryogenic Freezers

Cryogenic freezing is a relatively new method of freezing in which the food is exposed to an atmosphere below -60 °C through direct contact with liquefied gases such as nitrogen or carbon dioxide. This type of system differs from other freezing systems since it is not connected to a refrigeration plant; the refrigerants used are liquefied in large industrial installations and shipped to the food-freezing factory in pressure vessels. Thus, the small size and mobility of cryogenic freezers allow for flexibility in design and efficiency of the freezing application. Low initial investment and rather high operating costs are typical for cryogenic freezers.

Liquid Nitrogen Freezers

Liquid nitrogen, with a boiling temperature of -196 °C at atmospheric pressure, is a by-product of oxygen manufacture. The refrigerant is sprayed into the freezer and evaporates both on leaving the spray nozzles and on contact with the products. The system is designed in a way that the refrigerant passes in counter current to the movement of the products on the belt giving high transfer efficiency. The refrigerant consumption is in the range of 1.2-kg refrigerant per kg of the product. Typical food products used in this system are, fish fillets, seafood, fruits, berries.

Liquid Carbon Dioxide Freezers

Liquid carbon dioxide exists as either a solid or gas when stored at atmospheric pressure. When the gas is released to the atmosphere at -70 °C, half of the gas becomes dry-ice snow and the other half stays in the form of vapor. This unusual property of liquid carbon dioxide is used in a variety of freezing systems, one of which is a pre-freezing treatment before the product is exposed to nitrogen spray.

Packaging

Proper packaging of frozen food is important to protect the product from contamination and damage while in transit from the manufacturer to the consumer, as well as to preserve food value, flavour, colour, and texture. There are several factors considered in designing a suitable package for a frozen food. The package should be attractive to the consumer, protected from external contamination, and effective in terms of processing, handling, and cost. Proper selection is based on the type of package and material. There are typically three types of packaging used for frozen foods: primary, secondary, and tertiary. The primary package is in direct contact with the food and the food is kept inside the package up to the time of use. Secondary packaging is a form of multiple packaging used to handle packages together for sale. Tertiary packaging is used for bulk transportation of products.

Packaging materials should be moisture-vapor-proof to prevent evaporation, thus retaining the highest quality in frozen foods. Oxygen should also be completely evacuated from the package using a vacuum or gas-flush system to prevent migration of moisture and oxygen. Glass and rigid plastic are examples of moisture-vapor-proof packaging materials. Many packaging materials, however, are not moisture-vapor-proof, but are sufficiently moisture-vapor-resistant to retain satisfactory quality in foods. Most bags, wrapping materials, and waxed cartons used in freezing packaging are moisture-vapor-resistant. In general, the containers should be leakage free while easy to seal. Durability of the material is another important factor to consider, since the packaging material must not become brittle at low temperatures and crack (MSU, 1999).

A range of different packaging materials, mainly grouped as rigid and non-rigid containers, can be used for primary packaging. Glass, plastic, tin, and heavily waxed cardboard materials are in the rigid container group and usually used for packaging of liquid food products. Glass containers are mostly used for fruits and vegetables if they are not water-packed. Plastics are the derivatives of the oil-cracking industry (Brydson, 1982). Non-rigid containers include bags and sheets made of moisture-vapor-resistant heavy aluminum foil, polyethylene or laminated papers. Bags are the most commonly used packaging materials for frozen fruits and vegetables due to their flexibility during processing and handling. They can be used with or without outer cardboard cartons to protect against tearing.

Shape and size of the container are also important factors in freezing products. Serving size may vary depending on the type of product and selection should be based on the amount of food determined for one meal. For shape of the container, freezer space must be considered since rigid containers with flat tops and bottoms stack well in the freezer, while round containers waste freezer space.

Frozen Storage and Distribution

The quality of the final product depends on the history of the raw material. Using the lowest possible temperature is essential for frozen storage, transport, and distribution in achieving a high-quality product, since deteriorative processes are mainly temperature dependent. The lower the product temperature is, the slower the speed of reaction is leading to loss of quality. The temperatures of supply chains in freezing applications from the factory to the retail cabinet should be carefully monitored. The temperature regime covering the freezing process, the cold-store temperatures (£ -18 °C), distribution temperatures (£ -15 °C), and retail display (£ -12 °C) are given as legal standards.

Freezing Fruits and Vegetables in Small and Medium Scale Operations and its Potential Applications in Warm Climates

The preservation of fruits and vegetables by freezing is one of the most important methods for retaining high quality in agricultural products over long-term storage. In particular, the freshness qualities of raw fruits and vegetables can be retained for long periods, extending well beyond the normal season of most horticultural crops. The potential application of freezing preservation of fruits and vegetables, including tropical products, has been increasing recently in parallel with developments in developing countries.

Freezing Fruits

The effect of freezing, frozen storage, and thawing on fruit quality has been investigated over several decades. Today frozen fruits constitute a large and important food group (Skrede, 1996). The quality demanded in frozen fruit products is mostly based on the intended use of the product. If the fruit is to be eaten without any further processing after thawing, texture characteristics are more important when compared to use as a raw material in other industries. In general, conventional methods of freezing tend to destroy the turgidity of living cells in fruit tissue. Different from vegetables, fruits do not have a fibrous structure that can resist this destructive effect. Additionally, fruits to be frozen are harvested in a fully ripe state and are soft in texture. On the contrary, a great number of vegetables are frozen in an immature state (Boyle *et al.,* 1977). Fruits have delicate flavours that are easily damaged or changed by heat, indicating they are best eaten when raw and decrease in quality with processing. In the same way, attractive colour is important for frozen fruits. Chemical treatments or additives are often used to inactivate the deteriorative enzymes in fruits. Therefore, proper processing is essential for all steps involved, from harvesting to packaging and distribution.

Production and Harvesting

The characteristics of raw materials are of primary importance in determining the quality of the frozen product. These characteristics include several factors such as genetic makeup, climate of the growing area, type of fertilization, and maturity of harvest.

The ability to withstand rough handling, resistance to virus diseases, molds, uniformity in ripening, and yield are some of the important characteristics of fruits in terms of economical aspects considered in production. The use of mechanical harvesting generally causes bruising of fruits and results in a wide range of maturity levels for fruits. In contrast, hand-picking provides gentler handling and maturity sorting of fruits. However in most cases, it is non-economical compared to mechanical harvesting due to high labour cost.

As a rule, harvesting of fruits at an optimum level for commercial use is difficult. Simple tests like pressure tests are applied to determine when a fruit has reached optimum maturity for harvest. Colour is also one of the characteristics used in determining maturity since increased maturation causes a darker colour in fruits. A combination of colour and pressure tests is a better way to assess maturity level for harvesting.

Controlled atmosphere storage is a common method of storage for some fruits prior to freezing. In principle, a controlled atmosphere high in carbon dioxide and low in oxygen content slows down the rate of respiration, which may extend shelf life of any respiring fruit during storage. Due to the fact that these fruits do not ripen appreciably after picking, most fruits are picked as near to eating-ripe maturity as possible.

Pre-process Handling and Operations

Freezing preservation of fruits can only help retain the inherent quality present initially in a product since the process does not improve the quality characteristics of raw materials. Therefore, quality level of the raw materials prior to freezing is the major consideration for successful freezing. Washing and cutting generally results in losses when applied after thawing. Thus, fruits should be prepared prior to the freezing process in terms of peeling, slicing or cutting. Freezing preservation does not require specific unit operations for cleaning, rinsing, sorting, peeling, and cutting of fruits.

Fruits that require peeling before consumption should be peeled prior to freezing. Peeling is done by scalding the fruit in hot water, steam or hot lye solutions. The effect of peeling on the quality of frozen products has been studied for several fruits, including kiwi, banana, and mango. The rate of freezing can be increased by decreasing the size of products frozen, especially for large fruits. An increase in the freezing rate results in smaller ice crystals, which decreases cellular damage in fruit tissue. Banana, tomato, mango, and kiwi are some examples of large fruits commonly cut into smaller cubes or slices prior to freezing.

The objective of blanching is to inactivate the enzymes causing detrimental changes in colour, odour, flavour, and nutritive value, but heat treatment causes loss of such characteristics in fruits. Therefore, only a few types of fruits are blanched for inactivation of enzymes prior to freezing. The loss of water-soluble minerals and vitamins during blanching should also be minimized by keeping blanching time and temperature at an optimum combination.

Effect of Ingredients

Addition of sugars is an extremely important pretreatment for fruits prior to freezing since the treatment has the effect of excluding oxygen from the fruit, which helps to retain colour and appearance. Sugars when dissolved in solutions act by withdrawing water from cells by osmosis, resulting in very concentrated solutions inside the cells. The high concentration of solutes depresses the freezing point and therefore reduces the freezing within the cells, which inhibits excessive structural damage. Sugar syrups in the range of 30-60 percent sugar content are commonly used to cover the fruit completely, acting as a barrier to oxygen transmission and browning. Several experiments have shown the protective effect of sugar on flavour, odour, colour, and nutritive value during freezing, especially for frozen berries.

Packaging

Fruits exposed to oxygen are susceptible to oxidative degradation, resulting in browning and reduced storage life of products. Therefore, packaging of frozen fruits is based on excluding air from the fruit tissue. Replacement of oxygen with sugar solution or inert gas, consuming the oxygen by glucose-oxidase and/or the use of vacuum and oxygen-impermeable films are some of the methods currently employed for packaging frozen fruits. Plastic bags, plastic pots, paper bags, and cans are some of the most commonly used packaging materials (with or without oxygen removal) selected, based on penetration properties and thickness.

There are several types of fruit packs suitable for freezing: syrup pack, sugar pack, unsweetened pack, and tray pack and sugar replacement pack. The type of pack is usually selected according to the intended use for the fruit. Syrup-packed fruits are generally used for cooking purposes, while dry-packed and tray-packed fruits are good for serving raw in salads and garnishes.

Syrup Pack

The proportion of sugar to water used in a syrup pack depends on the sweetness of the fruit and the taste preference of the consumer. For most fruits, 40 percent sugar syrup is recommended. Lighter syrups are lower in calories and mostly desirable for mild-flavoured fruits to prevent masking the flavour, while heavier syrups may be used for very sour fruits.

Syrup is prepared by dissolving the sugar in warm water and cooling the solution down before usage. Just enough cooled syrup is used to cover the prepared fruit after it has been settled by jarring the container. In order to keep the fruit under the syrup, a small piece of crumpled waxed paper or other water resistant wrapping material is placed on top; the fruit is pressed down into the syrup before closing, then sealed and frozen.

Pectin can be used to reduce sugar content in syrups when freezing berries, cherries, and peaches. Pectin syrups are prepared by dissolving 1 box of powdered pectin with 1 cup of water. The solution is stirred and boiled for 1 minute; 1/2 cup of sugar is added and dissolved; the solution is then cooled down with the addition of cold water. Previously prepared fruit is put into a 4 to 6 quart bowl and enough pectin syrup is added to cover the fruit with a thin film. The pack is sealed and promptly frozen.

Sugar Packs

In preparing a sugar pack, sugar is first sprinkled over the fruit. Then the container is agitated gently until the juice is drawn out and the sugar is dissolved. This type of pack is generally used for soft sliced fruits such as peaches, strawberries, plums, and cherries, by using sufficient syrup to cover the fruit. Some whole fruits may also be coated with sugar prior to freezing (Beck, 1996).

Unsweetened Packs

Unsweetened packs can be prepared in several ways, either dry-packed, covered with water containing ascorbic acid, or packed in unsweetened juice. When water or juice is used in syrup and sugar packs, fruit is submerged by using a small piece of crumpled water-resistant material. Generally, unsweetened packs yield a lower quality product when compared with sugar packs, with the exception, some fruits such as raspberries, blueberries, scalded apples, gooseberries, currants, and cranberries maintain good quality without sugar (Beck, 1996).

Tray Packs

Unsweetened packs are generally prepared by using tray packs in which a single layer of prepared fruit is spread on shallow trays, frozen, and packaged in freezer bags promptly. The fruit sections remain loose without clumping together, which offers the advantage of using frozen fruit piece by piece.

Sugar Replacement Packs

Artificial sweeteners can be used instead of sugar in the form of sugar substitutes. The sweet taste of sugar can be replaced by using these kinds of sweeteners, however the beneficial effects of sugar like colour protection and thick syrup can not be replaced. Fruits frozen with sugar substitutes will freeze harder and thaw more slowly than fruits preserved with sugar.

Freezing Vegetables

Freezing is often considered the simplest and most natural way of preservation for vegetables (Cano, 1996). Frozen vegetables and potatoes form a significant proportion of the market in terms of frozen food consumption (Mallett, 1993). The quality of frozen vegetables depends on the quality of fresh products, since freezing does not improve product quality. Pre-process handling, from the time vegetables are picked until ready to eat, is one of the major concerns in quality retention.

***Table:** Fruit freezing guide*

Fruit	*Preparation*	*Type of Pack*
Apples	Wash, peel, and slice into antidarkening solution - 3 tablespoons lemon juice per quart of water	Pack in 30-40% syrup, adding 1/2 teaspoon crystalline ascorbic acid per quart of syrup. Pack dry or with up to 1/2 cup sugar per quart of apple slices.
Apricots	Wash, halve, and pit. Peel and slice if desired. If apricots are not peeled, heat in boiling water for 1/2 minute to keep skins from toughening during freezing. Cool in cold water, drain.	Pack in 40% syrup, adding 3/4 teaspoon crystalline ascorbic acid per quart of syrup.
Avocados	Peel soft, ripe avocados. Cut in half, remove pit, mash pulp.	Add 1/8 teaspoon crystalline ascorbic acid to each quart of puree. Package in recipe-size amounts.
Berries	Select firm, fully ripe berries. Sort, wash, and drain.	Use 30% syrup pack, dry unsweetened pack, dry sugar pack, (3/4 cup sugar per quart of berries), or tray pack.
Cherries (sour or sweet)	Select well-colored, tree-ripened cherries. Stem, sort, and wash thoroughly. Drain and pit.	Pack in 30-40% syrup. Add 1/2 teaspoon ascorbic acid per quart of syrup. For pies and other cooked products, pack in dry sugar using 3/4-cup sugar per quart of fruit.
Citrus fruits, (sections or slices)	Select firm fruit, free of soft spots. Wash and peel.	Pack in 40% syrup or in fruit juice. Add 1/2 teaspoon ascorbic acid per quart of syrup or juice.
Grapes	Select firm, ripe grapes. Wash and remove stems. Leave seedless grapes whole. Cut grapes with seeds in half and remove seeds.	Pack in 20% syrup or pack without sugar. Use dry pack for halved grapes and tray pack for whole grapes.
Melons (cantaloupe, watermelon)	Select firm-fleshed, well-colored, ripe melons. Wash rinds well. Slice or cut into chunks.	Pack in 30% syrup or pack dry using no sugar. Pulp also may be crushed (except watermelon), adding 1 tablespoon sugar per quart. Freeze in recipe-size containers.

Crop Cultivar, Production, and Maturity

The choice of the right cultivar and maturity before crop is harvested are the two most important factors affecting raw material quality. Raw material characteristics are usually related to the vegetable cultivar, crop production, crop maturity, harvesting practices, crop storage, transport, and factory reception.

The choice of crop cultivars is mostly based on their suitability for frozen preservation in terms of factory yield and product quality. Some of the characteristics used as selection criteria are as follows (Cano, 1996):

- Suitability for mechanical harvesting
- Uniform maturity
- Exceptional flavour and uniform colour and desirable texture
- Resistance to diseases
- High yield

Although cultivar selection is a major factor affecting the quality of the final product, many practices in the field and factors during growth of crop can also have a significant effect on quality. Those practices include site selection for growth, nutrition of crop, and use of agricultural chemicals to control pests or diseases. The maturity assessment for harvesting is one of the most difficult parts of the production. In addition to conventional methods, new instruments and tests have been developed to predict the maturity of crops that help determining the optimum harvest time, although the maturity assessment differs according to crop variety (Hui *et al.*, 2004).

Harvesting

At optimum maturity, physiological changes in several vegetables take place very rapidly. Thus, the determination of optimum harvesting time is critical (Arthey, 1993). Some vegetables such as green peas and sweet corn only have a short period during which they are of prime quality. If harvesting is delayed beyond this point, quality deteriorates and the crop may quickly become unacceptable (Lee, 1989). Most of the vegetables are subjected to bruising during harvesting.

Pre-process Handling

Vegetables at peak flavour and texture are used for freezing. Postharvest delays in handling vegetables are known to produce deterioration in flavour, texture, colour, and nutrients (Lee, 1989). Therefore, the delays between harvest and processing should be reduced to retain fresh quality prior to freezing. Cooling vegetables by cold water, air blasting, or ice will often reduce the rate of post-harvest losses sufficiently, providing extra hours of high quality retention for transporting raw material to considerable distances from the field to the processing plant (Deitrich *et al.,* 1977).

Blanching

Blanching is the exposure of the vegetables to boiling water or steam for a brief period of time to inactivate enzymes. Practically every vegetable (except herbs and green peppers) needs to be blanched and promptly cooled prior to freezing, since heating slows or stops the enzyme action, which causes vegetables to grow and mature. After maturation, however, enzymes can cause loss in quality, flavour, colour, texture, and nutrients. If vegetables are not heated sufficiently, the enzymes will continue to be active during frozen storage and may cause the vegetables to toughen or develop off-flavours and colours. Blanching also causes wilting or softening of vegetables, making them easier to pack. It destroys some bacteria and helps remove any surface dirt.

Blanching in hot water at 70 to 105 °C has been associated with the destruction of enzyme activity. Blanching is usually carried out between 75 and 95 °C for 1 to 10 minutes, depending on the size of individual vegetable pieces (Holdsworth, 1983). Blanched vegetables should be promptly cooled down to control and minimize the degradation of soluble and heat-labile nutrients.

The enzymes used as indicators of effectiveness of the blanching treatment are peroxidase, catalase, and more recently lipoxygenase. Peroxidase inactivation is commonly used in vegetable processing, since peroxidase is easily detected and is the most heat stable of these enzymes.

Vegetables can be blanched in hot water, steam, and in the microwave. Hot water blanching is the most common way of processing vegetables. Blanching times recommended for various vegetables, which indicates that the operation time can vary depending on the intended product use. For water blanching, vegetables are put in a basket and then placed in a kettle of boiling water covered with a lid. Timing begins immediately (Archuleta, 2003). Steam blanching takes longer than the water method, but helps retain water-soluble nutrients such as water-soluble vitamins. For steam blanching, a single layer of vegetables is placed on a rack or in a basket at 3-5 cm above water boiling in a kettle. A tightly fitted lid is placed on the kettle and timing is started. Microwave blanching is usually recommended for small quantities of vegetables prior to freezing. Due to the non-uniform heating disadvantage of microwaves, research is still being conducted to obtain better results with microwave blanching.

Table: *Vegetable freezing guide*

Vegetable	*Preparation*	*Blanch/Freeze*	
Asparagus	Wash and sort by size. Snap off tough ends. Cut stalks into 5-cm lengths.	Water blanch:	2 min
		Steam blanch:	3 min
Beans	Wash and trim the ends. Cut if desired.	Water blanch:	Steam blanch:
		Whole: 3 min.	Whole: 4 min.
		Cut: 2min.	Cut: 3min.
Beets	Wash and remove the tops leaving 2.5 cm of stem and root.	Cook until tender: 25-30 min Cool promptly, peel, trim. Cut into slices or cubes and pack.	
Broccoli	Wash and cut into pieces.	Water blanch:	3 min.
		Steam blanch:	3 min.
Cabbage	Wash and cut into wedges.	Water blanch:	3 min.
		Steam blanch:	4 min.
Carrots	Wash, peel and trim. Cut if desired.	Water blanch: 5 min.	
Cauliflower	Discard leaves; steam and wash. Break into flowerets.	Water blanch:	Steam blanch:
		Whole: 5 min.	Whole: 7 min
Corn	Remove husks and silks. Trim ends and wash.	Water blanch:	Steam blanch:
		Whole: 5 min.	Whole: 7 min
Greens	Select young tender greens. Wash and trim the leaves.	Water blanch:	2 min.
		Steam blanch:	3 min.
Herbs	Wash.	No heat treatment is needed.	
Mushrooms	Wipe and damp with paper towel. Trim hard tip of stems. Sort and cut large mushrooms.	May be frozen without heat treatment.	
Peas	Shell garden peas.	Water blanch:	Steam blanch:
		1-1/2 min.	1-1/2 min.
Peppers	Wash, remove stems and seeds.	Freeze whole or cut as desired. No heat treatment is needed.	
Potatoes	Peel, cut or grate as desired.	Water blanch:	
		Whole: 5 min.	
		Pieces: 2-3 min.	

Packaging

There are several factors to consider in packaging frozen vegetables, which include protection from atmospheric oxygen, prevention of moisture loss, retention of flavour, and rate of heat transfer through the package (Arthey, 1996). There are two basic packing methods recommended for frozen vegetables: dry pack and tray pack.

In the dry pack method, the blanched and drained vegetables are put into meal-sized freezer bags and packed tightly to cut down on the amount of air in the package. Proper headspace (approximately 2 cm) is left at the top of rigid containers before closing. For freezer bags, the headspace is larger. Provision for headspace is not necessary for foods such as broccoli, asparagus, and brussels sprouts, as they do not pack tightly in containers. In the tray pack method, chilled, well-drained vegetables are placed in a single layer on shallow trays or pans. Trays are placed in a freezer until the vegetables become firm, then removed. Vegetables are filled into containers. Tray-packed foods do not freeze in a block but remain loosely distributed so that the amount needed can be poured from the container and the package reclosed.

6

Dairy Process Engineering

A dairy is a business enterprise established for the harvesting of animal milk – mostly from cows or goats, but also from buffalo, sheep, horses or camels – for human consumption. A dairy is typically located on a dedicated *dairy farm* or section of a multi-purpose farm that is concerned with the harvesting of milk.

Terminology differs between countries. For example, in the United States, a farm building where milk is harvested is often called a "milking parlor". In New Zealand such a building is historically known as a "milking shed" or "milking parlour" (note the different spelling). Sometimes milking sheds are referred to by their type, such as "herring bone shed" or "pit parlour". In some countries, especially those with small numbers of animals being milked, as well as harvesting the milk from an animal, the dairy may also process the milk into butter, cheese and yogurt, for example. This is a traditional method of producing specialist milk products, especially in Europe. In the United States a *dairy* can also be a place that processes, distributes and sells dairy products, or a room, building or establishment where milk is stored and processed into milk products, such as butter or cheese. In New Zealand English the singular use of the word *dairy* almost exclusively refers to a corner shop, or superette. This usage is historical as such shops were a common place for the public to buy milk products.

As an attributive, the word *dairy* refers to milk-based products, veil, derivatives and processes, and the animals and workers involved in their production: for example dairy cattle, dairy goat. A dairy farm produces milk and a dairy factory processes it into a variety of dairy products. These establishments constitute the dairy industry, a component of the food industry.

As in many other branches of the food industry, dairy processing in the major dairy producing countries has become increasingly concentrated, with fewer but larger and more efficient plants operated by fewer workers. This is notably the case in the United States, Europe, Australia and New Zealand. In 2009, charges of anti-trust violations have been made against major dairy industry players in the United States.

Government intervention in milk markets was common in the 20th century. A limited anti-trust exemption was created for U.S. dairy cooperatives by the Capper-Volstead Act of 1922. In the 1930s, some U.S. states adopted price controls, and Federal Milk Marketing Orders started under the Agricultural Marketing Agreement Act of 1937 and continue in the 2000s. The Federal Milk Price Support Programme began in 1949. The Northeast Dairy Compact regulated wholesale milk prices in New England from 1997 to 2001.

Plants producing liquid milk and products with short shelf life, such as yogurts, creams and soft cheeses, tend to be located on the outskirts of urban centres close to consumer markets. Plants manufacturing items with longer shelf life, such as butter, milk powders, cheese and whey powders, tend to be situated in rural areas closer to the milk supply. Most large processing plants tend to specialise in a limited range of products. Exceptionally, however, large plants producing a wide range of products are still common in Eastern Europe, a holdover from the former centralized, supply-driven concept of the market.

As processing plants grow fewer and larger, they tend to acquire bigger, more automated and more efficient equipment. While this technological tendency keeps manufacturing costs lower, the need for long-distance transportation often increases the environmental impact.

Milk production is irregular, depending on cow biology. Producers must adjust the mix of milk which is sold in liquid form vs. processed foods (such as butter and cheese) depending on changing supply and demand.

Operation of the Dairy Farm

When it became necessary to milk larger cows, the cows would be brought to a shed or barn that was set up with bails (stalls) where the cows could be confined while they were milked. One person could milk more cows this way, as many as 20 for a skilled worker. But having cows standing about in the yard and shed waiting to be milked is not good for the cow, as she needs as much time in the paddock grazing as is possible. It is usual to restrict the twice-daily milking to a maximum of an hour and a half each time. It makes no difference whether one milks 10 or 1000 cows, the milking time should not exceed a total of about three hours each day for any cow.

As herd sizes increased there was more need to have efficient milking machines, sheds, milk-storage facilities (vats), bulk-milk transport and shed cleaning capabilities and the means of getting cows from paddock to shed and back.

Farmers found that cows would abandon their grazing area and walk towards the milking area when the time came for milking. This is not surprising as, in the flush of the milking season, cows presumably get very uncomfortable with udders engorged with milk, and the place of relief for them is the milking shed.

As herd numbers increased so did the problems of animal health. In New Zealand two approaches to this problem have been used. The first was improved veterinary medicines (and the government regulation of the medicines) that the farmer could use. The other was the creation of *veterinary clubs* where groups of farmers would employ a veterinarian (vet) full-time and share those services throughout the year. It was in the vet's interest to keep the animals healthy and reduce the number of calls from farmers, rather than to ensure that the farmer needed to call for service and pay regularly.

Most dairy farmers milk their cows with absolute regularity at a minimum of twice a day, with some high-producing herds milking up to four times a day to lessen the weight of large volumes of milk in the udder of the cow. This daily milking routine goes on for about 300 to 320 days per year that the cow stays in milk. Some small herds are milked once a day for about the last 20 days of the production cycle but this is not usual for large herds. If a cow is left unmilked just once she is likely to reduce milk-production almost immediately and the rest of the season may see her *dried off* (giving no milk) and still consuming feed for no production. However, once-a-day milking is now being practised more widely in New Zealand for profit and lifestyle reasons. This is effective because the fall in milk yield is at least partially offset by labour and cost savings from milking once per day. This compares to some intensive farm systems in the United States that milk three or more times per day due to higher milk yields per cow and lower marginal labour costs.

Farmers who are contracted to supply liquid milk for human consumption (as opposed to milk for processing into butter, cheese, and so on—see milk) often have to manage their herd so that the contracted number of cows are in milk the year round, or the required minimum milk output is maintained. This is done by mating cows outside their natural mating time so that the period when each cow in the herd is giving maximum production is in rotation throughout the year.

Northern hemisphere farmers who keep cows in barns almost all the year usually manage their herds to give continuous production of milk so that they get paid all year round. In the southern hemisphere the cooperative dairying systems allow for two months on no productivity because their systems are designed to take advantage of maximum grass and milk production in the spring and because the milk processing plants pay bonuses in the dry (winter) season to carry the farmers through the mid-winter break from milking. It also means that cows have a rest from milk production when they are most heavily pregnant. Some year-round milk farms are penalised financially for over-production at any time in the year by being unable to sell their overproduction at current prices.

Artificial insemination (AI) is common in all high-production herds.

Industrial Processing

Figure: *A Fonterra cooperative dairy factory in Australia.*

Dairy plants process the raw milk they receive from farmers so as to extend its marketable life. Two main types of processes are employed: heat treatment to ensure the safety of milk for human consumption and to lengthen its shelf-life, and dehydrating dairy products such as butter, hard cheese and milk powders so that they can be stored.

Cream and Butter

Today, milk is separated by huge machines in bulk into cream and skim milk. The cream is processed to produce various consumer products, depending on its thickness, its suitability for culinary uses and consumer demand, which differs from place to place and country to country.

Some cream is dried and powdered, some is condensed (by evaporation) mixed with varying amounts of sugar and canned. Most cream from New Zealand and Australian factories is made into butter. This is done by churning the cream until the fat globules coagulate and form a monolithic mass. This butter mass is washed and, sometimes, salted to improve keeping qualities. The residual buttermilk goes on to further processing. The butter is packaged (25 to 50 kg boxes) and chilled for storage and sale. At a later stage these packages are broken down into home-consumption sized packs.

Skimmed Milk

The product left after the cream is removed is called skim, or skimmed, milk.To make a consumable liquid a portion of cream is returned to the skim milk to make *low fat milk* (semi-skimmed) for human consumption. By varying the amount of cream returned, producers can make a variety of low-fat milks to suit their local market. Other products, such as calcium, vitamin D, and flavouring, are also added to appeal to consumers

Casein

Casein is the predominant phosphoprotein found in fresh milk. It has a very wide range of uses from being a filler for human foods, such as in ice cream, to the manufacture of products such as fabric, adhesives, and plastics.

Cheese

Cheese is another product made from milk. Whole milk is reacted to form curds that can be compressed, processed and stored to form cheese. In countries where milk is legally allowed to be processed without pasteurisation a wide range of cheeses can be made using the bacteria naturally in the milk. In most other countries, the range of cheeses is smaller and the use of artificial cheese curing is greater. Whey is also the byproduct of this process. Some people that are lactose intolerant can eat certain types of cheese.

Whey

In earlier times whey was considered to be a waste product and it was, mostly, fed to pigs as a convenient means of disposal. Beginning about 1950, and mostly since about 1980, lactose and many other products, mainly food additives, are made from both casein and cheese whey.

Yogurt

Yogurt (or yoghurt) making is a process similar to cheese making, only the process is arrested before the curd becomes very hard.

Milk Powders

Milk is also processed by various drying processes into powders. Whole milk, skim milk, buttermilk, and whey products are dried into a powder form and used for human and animal consumption. The main difference between production of powders for human or for animal consumption is in the protection of the process and the product from contamination. Some people drink milk reconstituted from powdered milk, because milk is about 88% water and it is much cheaper to transport the dried product.

Other Milk Products

Kumis is produced commercially in Central Asia. Although it is traditionally made from mare's milk, modern industrial variants may use cow's milk instead.

Milking

***Figure:** Preserved Express Dairies three-axle Milk Tank Wagon at the Didcot Railway Centre, based on an SR chassis*

Originally, milking and processing took place on the dairy farm itself. Later, cream was separated from the milk by machine on the farm, and transported to a factory to be made into butter. The skim milk was fed to pigs. This allowed for the high cost of transport (taking the smallest volume high-value product), primitive trucks and the poor quality of roads. Only farms close to factories could afford to take whole milk, which was essential for cheesemaking in industrial quantities, to them.

Originally milk was originally distributed in 'pails', a lidded bucket with a handle. These proved impractical for transport by road or rail, and so the milk churn was introduced, based on the tall conical shape of the butter churn. Later large railway containers, such as the British Railway Milk Tank Wagon were introduced, enabling the transport of larger quantities of milk, and over longer distances.

The development of refrigeration and better road transport, in the late 1950s, has meant that most farmers milk their cows and only temporarily store the milk in large refrigerated bulk tanks, from where it is later transported by truck to central processing facilities.

In many European countries, particularly the United Kingdom, milk is then delivered direct to customers' homes by a milk float.

Milking Machines

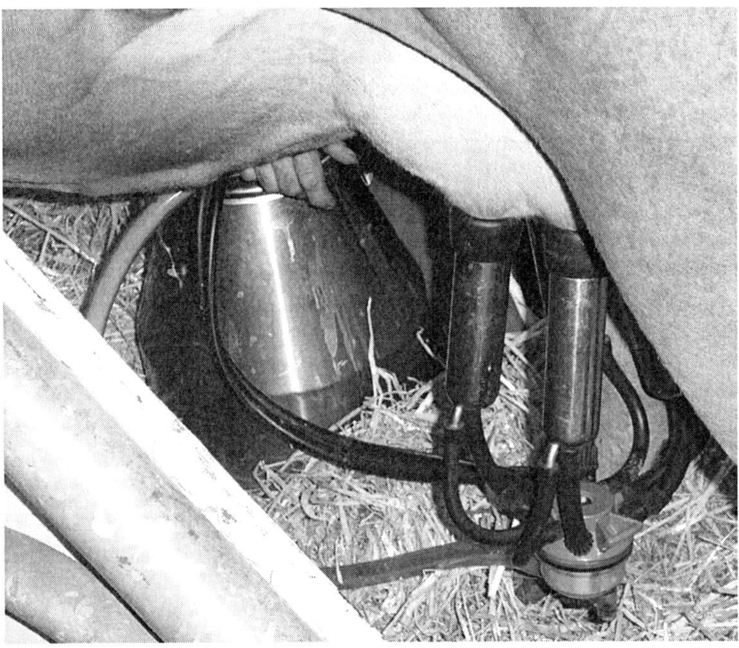

Figure: *The milking machine extracts milk from all teats.*

Milking machines are used to harvest milk from cows when manual milking becomes inefficient or labour intensive. One early model was patented in 1907. The milking unit is the portion of a milking machine for removing milk from an udder. It is made up of a claw, four teatcups, (Shells and rubber liners) long milk tube, long pulsation tube, and a pulsator. The claw is an assembly that connects the short pulse tubes and short milk tubes from the teatcups to the long pulse tube and long milk tube. (Cluster assembly) Claws are commonly made of stainless steel or plastic or both. Teatcups are composed of a rigid outer shell (stainless steel or plastic) that holds a soft inner liner or *inflation*. Transparent sections in the shell may allow viewing of liner collapse and milk flow. The annular space between the shell and liner is called the pulse chamber.

Milking machines work in a way that is different from hand milking or calf suckling. Continuous vacuum is applied inside the soft liner to massage milk from the teat by creating a pressure difference across the teat canal (or opening at the end of the teat). Vacuum also helps keep the machine attached to the cow. The vacuum applied to the teat causes congestion of teat tissues (accumulation of blood and other fluids). Atmospheric air is admitted into the pulsation chamber about once per second (the pulsation rate) to allow the liner to collapse around the end of teat and relieve congestion in the teat tissue. The ratio of the time that the liner is open (milking phase) and closed (rest phase) is called the pulsation ratio.

The four streams of milk from the teatcups are usually combined in the claw and transported to the milkline, or the collection bucket (usually sized to the output of one cow) in a single milk hose. Milk is then transported (manually in buckets) or with a combination of airflow and mechanical pump to a central storage vat or bulk tank. Milk is refrigerated on the farm in most countries either by passing through a heat-exchanger or in the bulk tank, or both.

The photo to the right shows a bucket milking system with the stainless steel bucket visible on the far side of the cow. The two rigid stainless steel teatcup shells applied to the front two quarters of the udder are visible. The top of the flexible liner is visible at the top of the shells as are the short milk tubes and short pulsation tubes extending from the bottom of the shells to the claw. The bottom of the claw is transparent to allow observation of milk flow. When milking is completed the vacuum to the milking unit is shut off and the teatcups are removed.

Milking machines keep the milk enclosed and safe from external contamination. The interior 'milk contact' surfaces of the machine are kept clean by a manual or automated washing procedures implemented after milking is completed. Milk contact surfaces must comply with regulations requiring food-grade materials (typically stainless steel and special plastics and rubber compounds) and are easily cleaned.

Most milking machines are powered by electricity but, in case of electrical failure, there can be an alternative means of motive power, often an internal combustion engine, for the vacuum and milk pumps.

Milking Shed Layouts

Figure: *Milking parlour at Pardes Hanna Agricultural High School, Israel*

Bail-style Sheds

This type of milking facility was the first development, after open-paddock milking, for many farmers. The building was a long, narrow, *lean-to* shed that was open along one long side. The cows were held in a yard at the open side and when they were about to be milked they were positioned in one of the bails (stalls). Usually the cows were restrained in the bail with a breech chain and a rope to restrain the outer back leg. The cow could not move about excessively and the milker could expect not to be kicked or trampled while sitting on a (three-legged) stool and milking into a bucket. When each cow was finished she backed out into the yard again. The UK bail, initially developed by Wiltshire dairy farmer Arthur Hosier, was a six standing mobile shed with steps that the cow mounted, so the herdsman didn't have to bend so low. The milking equipment was much as today, a vacuum from a pump, pulsators, a claw-piece with pipes leading to the four shells and liners that stimulate and suck the milk from the teat. The milk went into churns, via a cooler.

As herd sizes increased a door was set into the front of each bail so that when the milking was done for any cow the milker could, after undoing the leg-rope and with a remote link, open the door and allow her to exit to the pasture. The door was closed, the next cow walked into the bail and was secured. When milking machines were introduced bails were set in pairs so that a cow was being milked in one paired bail while the other could be prepared for milking. When one was finished the machine's cups are swapped to the other cow. This is the same as for *Swingover Milking Parlours* as described below except that the cups are loaded on the udder from the side. As herd numbers increased it was easier to double-up the cup-sets and milk both cows simultaneously than to increase the number of bails. About 50 cows an hour can be milked in a shed with 8 bales by one person. Using the same teat cups for successive cows has the danger of transmitting infection, mastitis, from one cow to another. Some farmers have devised their own ways to disinfect the clusters between cows.

Herringbone Milking Parlours

In herringbone milking sheds, or parlours, cows enter, in single file, and line up almost perpendicular to the central aisle of the milking parlour on both sides of a central pit in which the milker works (you can visualise a fishbone with the ribs representing the cows and the spine being the milker's working area; the cows face outward). After washing the udder and teats the cups of the milking machine are applied to the cows, from the rear of their hind legs, on both sides of the working area. Large herringbone sheds can milk up to 600 cows efficiently with two people.

Swingover Milking Parlours

Swingover parlours are the same as herringbone parlours except they have only one set of milking cups to be shared between the two rows of cows, as one side is being milked the cows on the other side are moved out and replaced with unmilked ones. The advantage of this system is that it is less costly to equip, however it operates at slightly better than half-speed and one would not normally try to milk more than about 100 cows with one person.

Rotary Milking Sheds

Rotary milking sheds (also known as Rotary milking parlor) consist of a turntable with about 12 to 100 individual stalls for cows around the outer edge. A "good" rotary will be operated with 24–32 (~48–50+) stalls by one (two) milkers. The turntable is turned by an electric-motor drive at a rate that one turn is the time for a cow to be milked completely. As an empty stall passes the entrance a cow steps on, facing the centre, and rotates with the turntable. The next cow moves into the next vacant stall and so on. The operator, or milker, cleans the teats, attaches the cups and does any other feeding or whatever husbanding operations that are necessary. Cows are milked as the platform rotates. The milker, or an automatic device, removes the milking machine cups and the cow backs out and leaves at an exit just before the entrance. The rotary system is capable of milking very large herds—over a thousand cows.

Automatic Milking Sheds

Automatic milking or 'robotic milking' sheds can be seen in Australia, New Zealand and many European countries. Current automatic milking sheds use the voluntary milking (VM) method. These allow the cows to voluntarily present themselves for milking at any time of the day or night, although repeat visits may be limited by the farmer through computer software. A robot arm is used to clean teats and apply milking equipment, while automated gates direct cow traffic, eliminating the need for the farmer to be present during the process. The entire process is computer controlled.

Supplementary Accessories in Sheds

Farmers soon realised that a milking shed was a good place to feed cows supplementary foods that overcame local dietary deficiencies or added to the cows' wellbeing and production. Each bail might have a box into which such feed is delivered as the cow arrives so that she is eating while being milked. A computer can read the eartag of each animal to ration the correct individual supplement. A close alternative is to use 'out-of-parlour-feeders', stalls that respond to a transponder around the cow's neck that is programmed to provide each cow with a supplementary feed, the quantity dependent on her production, stage in lactation, and the benefits of the main ration

The holding yard at the entrance of the shed is important as a means of keeping cows moving into the shed. Most yards have a powered gate that ensures that the cows are kept close to the shed.

Water is a vital commodity on a dairy farm: cows drink about 20 gallons (80 litres) a day, sheds need water to cool and clean them. Pumps and reservoirs are common at milking facilities. Water can be warmed by heat transfer with milk.

Temporary Milk Storage

Milk coming from the cow is transported to a nearby storage vessel by the airflow leaking around the cups on the cow or by a special "air inlet" (5-10 l/min free air) in the claw. From there it is pumped by a mechanical pump and cooled by a heat exchanger. The milk is then stored in a large vat, or bulk tank, which is usually refrigerated until collection for processing.

Waste Disposal

Figure: *Manure spreader going to the field from a dairy farm, Elba, New York.*

In countries where cows are grazed outside year-round, there is little waste disposal to deal with. The most concentrated waste is at the milking shed, where the animal waste is liquefied (during the water-washing process) and allowed to flow by gravity, or pumped, into composting ponds with anaerobic bacteria to consume the solids. The processed water and nutrients are then pumped back onto the pasture as irrigation and fertilizer.

In areas where cows are housed all year round, the waste problem is difficult because of the amount of feed that is brought in and the amount of bedding material that also has to be removed and composted.

In the associated milk processing factories, most of the waste is washing water that is treated, usually by composting, and returned to waterways. This is much different from half a century ago, when the main products were butter, cheese and casein, and the rest of the milk had to be disposed of as waste (sometimes as animal feed).

In many cases, modern farms have very large quantities of milk to be transported to a factory for processing. If anything goes wrong with the milking, transport or processing facilities it can be a major disaster trying to dispose of enormous quantities of milk. If a road tanker overturns on a road, the rescue crew is looking at accommodating the spill of 5 to 10 thousand gallons of milk (20 to 45 thousand litres) without allowing any into the waterways. A derailed rail tanker-train may involve 10 times that amount. Without refrigeration, milk is a fragile commodity, and it is very damaging to the environment in its raw state due to its high biochemical oxygen demand. A widespread electrical power blackout is another disaster for the dairy industry, because both milking and processing facilities are affected. For this, farms may often use mobile generators. Such a situation occurred during the power outage caused by the 2010 Canterbury Earthquake.

In dairy-intensive areas, various methods have been proposed for disposing of large quantities of milk. These directives include feeding milk to livestock, spray irrigation or designating a sacrifice area. Large application rates of milk onto land, or disposing in a hole, is problematic as the residue from the decomposing milk will block the soil pores and thereby reduce the water infiltration rate through the soil profile. As recovery of this effect can take time, any land based application needs to be well managed and considered.

Engineering Properties of Foods

Food is any substance consumed to provide nutritional support for the body. It is usually of plant or animal origin, and contains essential nutrients, such as carbohydrates, fats, proteins, vitamins, or minerals. The substance is ingested by an organism and assimilated by the organism's cells in an effort to produce energy, maintain life, or stimulate growth.

Historically, people secured food through two methods: hunting and gathering, and agriculture. Today, most of the food energy consumed by the world population is supplied by the food industry.

Food safety and food security are monitored by agencies like the International Association for Food Protection, World Resources

Institute, World Food Programme, Food and Agriculture Organization, and International Food Information Council. They address issues such as sustainability, biological diversity, climate change, nutritional economics, population growth, water supply, and access to food.

Physical properties of foods (including thermal, mechanical, rheological, dielectric, barrier properties, and water activity) are important for the proper design of food processing, handling, and storage systems. Various food processing methods can potential alter those physical properties and cause desirable or sometimes no so desirable changes in nutrient profiles, texture, colour, taste, aroma and other quality attributes. Therefore, it is important to investigate the physical and the impacted chemical properties of foods to gain insights on how they affect the quality attributes.

Food Industry

The food industry is a complex, global collective of diverse businesses that supply much of the food energy consumed by the world population. Only subsistence farmers, those who survive on what they grow, can be considered outside of the scope of the modern food industry.

The food industry includes:

- Regulation: local, regional, national and international rules and regulations for food production and sale, including food quality and food safety, and industry lobbying activities
- Education: academic, vocational, consultancy
- Research and development: food technology
- Financial services insurance, credit
- Manufacturing: agrichemicals, seed, farm machinery and supplies, agricultural construction, etc.
- Agriculture: raising of crops and livestock, seafood
- Food processing: preparation of fresh products for market, manufacture of prepared food products
- Marketing: promotion of generic products (e.g. milk board), new products, public opinion, through advertising, packaging, public relations, et
- Wholesale and distribution: warehousing, transportation, logistics

Industry Size

Processed food sales worldwide are approximately US$3.2 trillion (2004).

In 2012, U.S. consumers spent approximately US$1.8 trillion annually on food, or nearly 10 percent of the Gross Domestic Product (GDP). Over 16.5 million people are employed in the food industry.

In the United Kingdom, the food industry is extensive. It employs 2 million people and has a turnover in excess of £70bn. It is the largest manufacturing sector in the UK and represents around 15% of the total manufacturing sector in the UK. Around 13% of the people working in manufacturing in the UK work in the food and drink industry.

Agriculture

Agriculture is the process of producing food, feeding products, fibre and other desired products by the cultivation of certain plants and the raising of domesticated animals (livestock). The practice of agriculture is also known as "farming". Scientists, inventors and others devoted to improving farming methods and implements are also said to be engaged in agriculture. More people in the world are involved in agriculture as their primary economic activity than in any other, yet it only accounts for twenty percent of the world's Gross Domestic Product (GDP).

Food Processing

Figure: *Industrial cheese production*

Food processing is the transformation of raw ingredients into food, or of food into other forms. Food processing typically takes clean, harvested crops or butchered animal products and uses these to produce attractive, marketable and often long shelf-life food products.

Food processing dates back to the prehistoric ages when crude processing incorporated slaughtering, fermenting, sun drying, preserving with salt, and various types of cooking (such as roasting, smoking, steaming, and oven baking). Salt-preservation was especially common for foods that constituted warrior and sailors' diets until the introduction of canning methods. Evidence for the existence of these methods can be found in the writings of the ancient Greek, Chaldean, Egyptian and Roman civilizations as well as archaeological evidence from Europe, North and South America and Asia. These tried and tested processing techniques remained essentially the same until the advent of the industrial revolution. Examples of ready-meals also date back to before the preindustrial revolution, and include dishes such as Cornish pasty and Haggis. Both during ancient times and today in modern society these are considered processed foods. Food processing can provide quick, nutritious meal options for busy families.

Modern food processing technology developed in the 19th and 20th centuries was developed in a large part to serve military needs. In 1809 Nicolas Appert invented a hermetic bottling technique that would preserve food for French troops which ultimately contributed to the development of tinning, and subsequently canning by Peter Durand in 1810. Although initially expensive and somewhat hazardous due to the lead used in cans, canned goods would later become a staple around the world. Pasteurization, discovered by Louis Pasteur in 1864, improved the quality of preserved foods and introduced the wine, beer, and milk preservation.

In the 20th century, World War II, the space race and the rising consumer society in developed countries (including the United States) contributed to the growth of food processing with such advances as spray drying, juice concentrates, freeze drying and the introduction of artificial sweeteners, colouring agents, and preservatives such as sodium benzoate. In the late 20th century products such as dried instant soups, reconstituted fruits and juices, and self cooking meals such as MRE food ration were developed.

In western Europe and North America, the second half of the 20th century witnessed a rise in the pursuit of convenience. Food processing companies marketed their products especially towards middle-class working wives and mothers. Frozen foods (often credited to Clarence Birdseye) found their success in sales of juice concentrates and "TV dinners". Processors utilised the perceived value of time to appeal to the postwar population, and this same appeal contributes to the success of convenience foods today.

Benefits and Drawbacks

Benefits:

Figure: *Processed seafood - fish, squid, prawn balls and simulated crab sticks (surimi)*

Benefits of food processing include toxin removal, preservation, easing marketing and distribution tasks, and increasing food consistency. In addition, it increases yearly availability of many foods, enables transportation of delicate perishable foods across long distances and makes many kinds of foods safe to eat by de-activating spoilage and pathogenic micro-organisms. Modern supermarkets would not exist without modern food processing techniques, and long voyages would not be possible.

Processed foods are usually less susceptible to early spoilage than fresh foods and are better suited for long distance transportation from the source to the consumer. When they were first introduced, some processed foods helped to alleviate food shortages and improved the overall nutrition of populations as it made many new foods available to the masses.

Processing can also reduce the incidence of food borne disease. Fresh materials, such as fresh produce and raw meats, are more likely to harbour pathogenic micro-organisms (e.g. Salmonella) capable of causing serious illnesses. The extremely varied modern diet is only truly possible on a wide scale because of food processing. Transportation of more exotic foods, as well as the elimination of much hard labour gives the modern eater easy access to a wide variety of food unimaginable to their ancestors.

The act of processing can often improve the taste of food significantly.

Mass production of food is much cheaper overall than individual production of meals from raw ingredients. Therefore, a large profit potential exists for the manufacturers and suppliers of processed food products. Individuals may see a benefit in convenience, but rarely see any direct financial cost benefit in using processed food as compared to home preparation.

Processed food freed people from the large amount of time involved in preparing and cooking "natural" unprocessed foods. The increase in free time allows people much more choice in life style than previously allowed. In many families the adults are working away from home and therefore there is little time for the preparation of food based on fresh ingredients. The food industry offers products that fulfill many different needs: eg fully prepared ready meals that can be heated up in the microwave oven within a few minutes.

Modern food processing also improves the quality of life for people with allergies, diabetics, and other people who cannot consume some common food elements. Food processing can also add extra nutrients such as vitamins.

Drawbacks

Any processing of food can affect its nutritional density, the amount of nutrients lost depending on the food and method of processing. Vitamin C, for example, is destroyed by heat and therefore canned fruits have a lower content of vitamin C than fresh ones. The USDA conducted a study in 2004, creating a nutrient retention table for several foods. A cursory glance of the table indicates that, in the majority of foods, processing reduces nutrients by a minimal amount. On average any given nutrient may be reduced by as little as 5%-20%.

New research highlighting the importance to human health of a rich microbial environment in the intestine indicates that much food processing (but not fermentation of foods) endangers that balance.

Another safety concern in food processing is the use of food additives. The health risks of any additives will vary greatly from person to person; for example sugar as an additive would be detrimental to those with diabetes. In the European Union, only food additives (e.g., sweeteners, preservatives, stabilizers) that have been approved as safe for human consumption by the European Food Safety Authority (EFSA) are allowed, at specified levels, for use in food products. Approved additives receive an E number (E for Europe), which at the same time simplifies communication about food additives in the list of ingredients across the different languages of the EU.

Food processing is typically a mechanical process that utilizes large mixing, grinding, chopping and emulsifying equipment in the production process. These processes inherently introduce a number of contamination risks. As a mixing bowl or grinder is used over time the food contact parts will tend to fail and fracture. This type of failure will introduce into the product stream small to large metal contaminants. Further processing of these metal fragments will result in downstream equipment failure and the risk of ingestion by the consumer. Food manufacturers utilize industrial metal detectors to detect and reject automatically any metal fragment. Large food processors will utilize many metal detectors within the processing stream, to reduce both damage to processing machinery as well as risk to the consumer.

Performance Parameters for Food Processing

Figure: *Factory automation - robotics palettizing bread*

When designing processes for the food industry the following performance parameters may be taken into account:

- Hygiene, e.g. measured by number of micro-organisms per mL of finished product
- Energy efficiency measured e.g. by "ton of steam per ton of sugar produced"
- Minimization of waste, measured e.g. by "percentage of peeling loss during the peeling of potatoes"
- Labour used, measured e.g. by "number of working hours per ton of finished product"
- Minimization of cleaning stops measured e.g. by "number of hours between cleaning stops"

De-agglomerating Batter Mixes in Food Processing

Problems often occur during preparation of batter mixes because flour and other powdered ingredients tend to form lumps or agglomerates as they are being mixed during production. A conventional mixer/agitator cannot break down these agglomerates, resulting in a lumpy batter. If lumpy batter is used to enrobe products, it causes an unsatisfactory appearance with misshapen or oversize products that do not fit properly into packaging. This can force production to a standstill. Furthermore batter mix is generally recirculated from an enrobing system back to a holding vessel; lumps then have a tendency to build up, reducing the flow of material and raising potential sanitation issues.

Using a high shear in-line mixer in place of a conventional agitator or mixer can quickly solve problems of agglomeration with dry ingredients. A single pass through a self-pumping, in-line mixer adds high shear to batter, which de-agglomerates the mix, resulting in a homogeneous, smooth batter. With a consistent, smooth batter, finished product appearance is improved; the effectiveness and hygiene of the recirculation system is increased; and a better yield of raw materials is achieved. By increasing overall product quality, the amount of raw materials needed is decreased, thereby lowering manufacturing costs.

High shear in-line mixers process food to be made faster and cheaper while increasing consistency of the finished food. Powder and liquid mixing systems are capable of rapidly incorporating large quantities of powders at high concentrations – agglomerate free and fully hydrated. Advances in technology have made processing equipment easy to clean, leading to a much safer processed food.

Trends in Modern Food Processing

Health:

- Reduction of fat content in final product by using baking instead of deep-frying in the production of potato chips, another processed food.
- Maintaining the natural taste of the product by using less artificial sweetener than was used before.

Hygiene:

The rigorous application of industry and government endorsed standards to minimise possible risk and hazards. The international standard adopted is HACCP.

Efficiency:

- Rising energy costs lead to increasing usage of energy-saving technologies, e.g. frequency converters on electrical drives, heat insulation of factory buildings and heated vessels, energy recovery systems, keeping a single fish frozen all the way from China to Switzerland.
- Factory automation systems (often Distributed control systems) reduce personnel costs and may lead to more stable production results.

Industries: Food processing industries and practices include the following:

Cannery

Canning is a method of preserving food in which the food contents are processed and sealed in an airtight container. Canning provides a shelf life typically ranging from one to five years, although under specific circumstances a freeze-dried canned product, such as canned dried lentils, could last as long as 30 years in an edible state.

In 1795 the French military offered a cash prize of 12,000 francs for a new method to preserve food. Nicolas Appert suggested canning, and the process was first proven in 1806 in tests conducted by the French navy. Appert was awarded the prize in 1810 by Count Montelivert, a French minister of the interior. The packaging prevents microorganisms from entering and proliferating inside.

To prevent the food from being spoiled before and during containment, a number of methods are used: pasteurisation, boiling (and other applications of high temperature over a period of time), refrigeration, freezing, drying, vacuum treatment, antimicrobial agents that are natural to the recipe of the foods being preserved, a sufficient dose of ionizing radiation, submersion in a strong saline solution, acid, base, osmotically extreme (for example very sugary) or other microbially-challenging environments.

Other than sterilization, no method is perfectly dependable as a preservative. For example, the microorganism *Clostridium botulinum* (which causes botulism), can only be eliminated at temperatures above the boiling point.

From a public safety point of view, foods with low acidity (a pH more than 4.6) need sterilization under high temperature (116-130 °C). To achieve temperatures above the boiling point requires the use of a pressure canner. Foods that must be pressure canned include most vegetables, meat, seafood, poultry, and dairy products. The only foods that may be safely canned in an ordinary boiling water bath are highly acidic ones with a pH below 4.6, such as fruits, pickled vegetables, or other foods to which acidic additives have been added.

7

Fish and Meat Processing

The term fish processing refers to the processes associated with fish and fish products between the time fish are caught or harvested, and the time the final product is delivered to the customer. Although the term refers specifically to fish, in practice it is extended to cover any aquatic organisms harvested for commercial purposes, whether caught in wild fisheries or harvested from aquaculture or fish farming.

Larger fish processing companies often operate their own fishing fleets or farming operations. The products of the fish industry are usually sold to grocery chains or to intermediaries. Fish are highly perishable. A central concern of fish processing is to prevent fish from deteriorating, and this remains an underlying concern during other processing operations.

Fish processing can be subdivided into fish handling, which is the preliminary processing of raw fish, and the manufacture of fish products. Another natural subdivision is into primary processing involved in the filleting and freezing of fresh fish for onward distribution to fresh fish retail and catering outlets, and the secondary processing that produces chilled, frozen and canned products for the retail and catering trades.

There is evidence humans have been processing fish since the early Holocene. These days, fish processing is undertaken by artisan fishermen, on board fishing or fish processing vessels, and at fish processing plants.

Handling the Catch

When fish are captured or harvested for commercial purposes, they need some preprocessing so they can be delivered to the next part of the marketing chain in a fresh and undamaged condition. This means, for example, that fish caught by a fishing vessel need handling so they can be stored safely until the boat lands the fish on shore. Typical handling processes are

- transferring the catch from the fishing gear (such as a trawl, net or fishing line) to the fishing vessel
- holding the catch before further handling
- sorting and grading
- bleeding, gutting and washing
- chilling
- storing the chilled fish
- unloading, or landing the fish when the fishing vessel returns to port

The number and order in which these operations are undertaken varies with the fish species and the type of fishing gear used to catch it, as well as how large the fishing vessel is and how long it is at sea, and the nature of the market it is supplying. Catch processing operations can be manual or automated. The equipment and procedures in modern industrial fisheries are designed to reduce the rough handling of fish, heavy manual lifting and unsuitable working positions which might result in injuries.

Handling Live Fish

An alternative, and obvious way of keeping fish fresh is to keep them alive until they are delivered to the buyer or ready to be eaten. This is a common practice worldwide. Typically, the fish are placed in a container with clean water, and dead, damaged or sick fish are removed. The water temperature is then lowered and the fish are starved to reduce their metabolic rate. This decreases fouling of water with metabolic products (ammonia, nitrite and carbon dioxide) that become toxic and make it difficult for the fish to extract oxygen.

Fish can be kept alive in floating cages, wells and fish ponds. In aquaculture, holding basins are used where the water is continuously filtered and its temperature and oxygen level are controlled. In China, floating cages are constructed in rivers out of palm woven baskets, while in South America simple fish yards are built in the backwaters of rivers. Live fish can be transported by methods which range from simple artisanal methods where fish are placed in plastic bags with an oxygenated atmosphere, to sophisticated systems which use trucks that filter and recycle the water, and add oxygen and regulate temperature.

Preservation

Preservation techniques are needed to prevent fish spoilage and lengthen shelf life. They are designed to inhibit the activity of spoilage bacteria and the metabolic changes that result in the loss of fish quality. Spoilage bacteria are the specific bacteria that produce the unpleasant odours and flavours associated with spoiled fish. Fish normally host many bacteria that are not spoilage bacteria, and most of the bacteria present on spoiled fish played no role in the spoilage. To flourish, bacteria need the right temperature, sufficient water and oxygen, and surroundings that are not too acidic. Preservation techniques work by interrupting one or more of these needs. Preservation techniques can be classified as follows.

Control of Temperature

Figure: *Ice preserves fish and extends shelf life by lowering the temperature*

If the temperature is decreased, the metabolic activity in the fish from microbial or autolytic processes can be reduced or stopped. This is achieved by refrigeration where the temperature is dropped to about 0 °C, or freezing where the temperature is dropped below -18°C. On fishing vessels, the fish are refrigerated mechanically by circulating cold air or by packing the fish in boxes with ice. Forage fish, which are often caught in large numbers, are usually chilled with refrigerated or chilled seawater. Once chilled or frozen, the fish need further cooling to maintain the low temperature. There are key issues with fish cold store design and management, such as how large and energy efficient they are, and the way they are insulated and palletized.

An effective method of preserving the freshness of fish is to chill with ice by distributing ice uniformly around the fish. It is a safe cooling method that keeps the fish moist and in an easily stored form suitable for transport. It has become widely used since the development of mechanical refrigeration, which makes ice easy and cheap to produce. Ice is produced in various shapes; crushed ice and ice flakes, plates, tubes and blocks are commonly used to cool fish. Particularly effective is slurry ice, made from micro crystals of ice formed and suspended within a solution of water and a freezing point depressant, such as common salt.

A more recent development is pumpable ice technology. Pumpable ice flows like water, and because it is homogeneous, it cools fish faster than fresh water solid ice methods and eliminates freeze burns. It complies with HACCP and ISO food safety and public health standards, and uses less energy than conventional fresh water solid ice technologies.

Control of Water Activity

The water activity, a_w, in a fish is defined as the ratio of the water vapour pressure in the flesh of the fish to the vapour pressure of pure water at the same temperature and pressure. It ranges between 0 and 1, and is a parameter that measures how available the water is in the flesh of the fish. Available water is necessary for the microbial and enzymatic reactions involved in spoilage. There are a number of techniques that have been or are used to tie up the available water or remove it by reducing the a_w. Traditionally, techniques such as drying, salting and smoking have been used, and have been used for thousands of years. These techniques can be very simple, for example, by using solar drying. In more recent times, freeze-drying, water binding humectants, and fully automated equipment with temperature and humidity control have been added. Often a combination of these techniques is used.

Physical Control of Microbial Loads

Heat or ionizing irradiation can be used to kill the bacteria that cause decomposition. Heat is applied by cooking, blanching or microwave heating in a manner that pasteurizes or sterilizes fish products. Cooking or pasteurizing does not completely inactivate microorganisms and may need to be followed with refrigeration to preserve fish products and increase their shelf life. Sterilised products are stable at ambient temperatures up to 40°C, but to ensure they remain sterilized they need packaging in metal cans or retortable pouches before the heat treatment.

Chemical Control of Microbial Loads

Microbial growth and proliferation can be inhibited by a technique called biopreservation. Biopreservation is achieved by adding antimicrobials or by increasing the acidity of the fish muscle. Most bacteria stop multiplying when the pH is less than 4.5. Acidity is increased by fermentation, marination or by directly adding acids (acetic, citric, lactic) to fish products. Lactic acid bacteria produce the antimicrobial nisin which further enhances preservation. Other preservatives include nitrites, sulphites, sorbates, benzoates and essential oils.

Control of the Oxygen Reduction Potential

Spoilage bacteria and lipid oxidation usually need oxygen, so reducing the oxygen around fish can increase shelf life. This is done by controlling or modifying the atmosphere around the fish, or by vacuum packaging. Controlled or modified atmospheres have specific combinations of oxygen, carbon dioxide and nitrogen, and the method is often combined with refrigeration for more effective fish preservation.

Combined Techniques

Two or more of these techniques are often combined. This can improve preservation and reduce unwanted side effects such as the denaturation of nutrients by severe heat treatments. Common combinations are salting/drying, salting/marinating, salting/smoking, drying/smoking, pasteurization/refrigeration and controlled atmosphere/refrigeration. Other process combinations are currently being developed along the multiple hurdle theory.

Automated Processes

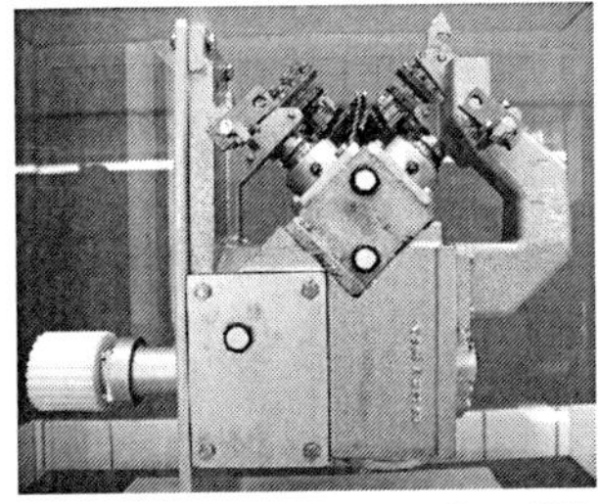

Figure: *Automatic knives for filleting fish*

"The search for higher productivity and the increase of labour cost has driven the development of computer vision technology, electronic scales and automatic skinning and filleting machines."

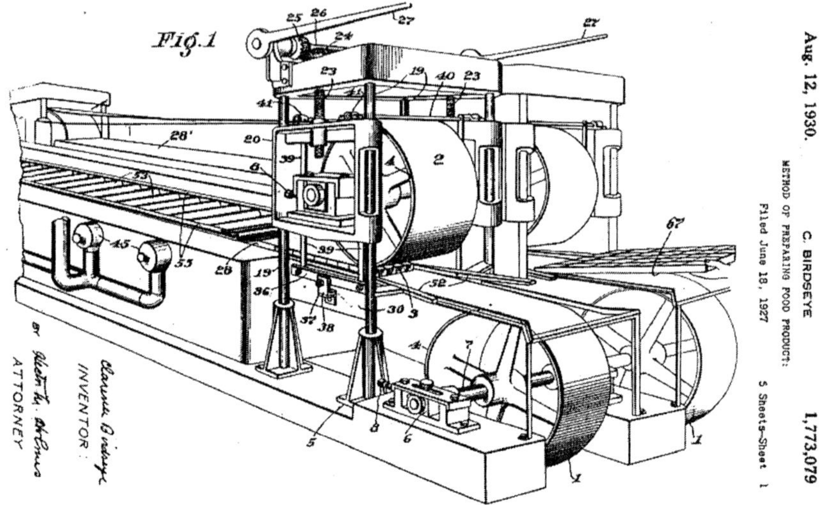

Figure: *Patent issued to Clarence Birdseye for the production of quick-frozen fish, 1930*

Figure: *Processing line for fish fingers*

Waste Management

Waste produced during fish processing operations can be solid or liquid.

- Solid wastes: include skin, viscera, fish heads and carcasses (fish bones). Solid waste can be recycled in fish meal plants or it can be treated as municipal waste.
- Liquid wastes: include bloodwater and brine from drained storage tanks, and water discharges from washing and cleaning. This waste may need holding temporarily, and should be disposed of without damage to the environment. How liquid waste should be disposed from fish processing operations depends on the content levels in the waste of solid and organic matter, as well as nitrogen and phosphorus content, and oil and grease content. It also depends on an assessment of parameters such acidity levels, temperature, odour, and biochemical oxygen demand and chemical oxygen demand. The magnitude of waste management issues depends on how much waste volume there is, the nature of the pollutants it carries, the rate at which it is discharged and the capacity of the receiving environment to assimilate the pollutants. Many countries dispose of such liquid wastes through their municipal sewage systems or directly into a waterway. The receiving waterbody should be able to degrade the organic and inorganic waste components in a way that does not damage the aquatic ecosystem.

Treatments can be primary and secondary.

- Primary treatments: use physical methods such as flotation, screening, and sedimentation to remove oil and grease and other suspended solids.
- Secondary treatments: use biological and physicochemical means. Biological treatments use microorganisms to metabolise the organic polluting matter into energy and biomass. "These microorganisms can be aerobic or anaerobic. The most used aerobic processes are activated sludge system, aerated lagoons, trickling filters or bacterial beds and the rotating biological contractors. In anaerobic processes, the anaerobic microorganisms digest the organic matter in tanks to produce gases (mainly methane and CO2) and biomass. Anaerobic digesters are sometimes heated, using part of the methane produced, to maintain a temperature of 30 to 35°C. In the physicochemical treatments, also called coagulation-flocculation, a chemical substance is added to the effluent to reduce the surface charges responsible for particle repulsions in a colloidal suspension, thus reducing the forces that keep

its particles apart. This reduction in charge causes flocculation (agglomeration) and particles of larger sizes are settled and clarified effluent is obtained. The sludge produced by primary and secondary treatments is further processed in digesting tanks through anaerobic processes or sprayed over land as a fertilizer. In the latter case, care must be exercised to ensure that the sludge is freed of its pathogens."

Transport

Fish is transported widely in ships, and by land and air, and much fish is traded internationally. It is traded live, fresh, frozen, cured and canned. Live, fresh and frozen fish need special care.

- Live fish: When live fish are transported they need oxygen, and the carbon dioxide and ammonia that result from respiration must not be allowed to build up. Most fish transported live are placed in water supersaturated with oxygen (though catfish can breathe air directly through their gills and body skin, and the climbing perch has special air-breathing organs). The fish are often "conditioned" (starved) before they are transported to reduce their metabolism and increase packing density, and the water can be cooled to further reduce metabolism. Live crustaceans can be packed in wet sawdust to keep the air humid.
- By air: Over five percent of the global fish production is transported by air. Air transport needs special care in preparation and handling and careful scheduling. Airline transport hubs often require cargo transfers under their own tight schedules. This can influence when the product is delivered, and consequently the condition it is in when it is delivered. The air shipment of leaking seafood packages causes corrosion damage to aircraft, and each year, in the US, requires millions of dollars to repair the damage. Most airlines prefer fish that is packed in dry ice or gel, and not packed in ice.
- By land or sea: "The most challenging aspect of fish transportation by sea or by road is the maintenance of the cold chain, for fresh, chilled and frozen products and the optimisation of the packing and stowage density. Maintaining the cold chain requires the use of insulated containers or transport vehicles and adequate quantities of coolants or mechanical refrigeration. Continuous temperature monitors are used to provide evidence that the cold chain has not been broken during transportation. Excellent development in food packaging and handling allow rapid and

efficient loading, transport and unloading of fish and fishery products by road or by sea. Also, transport of fish by sea allows for the use of special containers that carry fish under vacuum, modified or controlled atmosphere, combined with refrigeration."

Quality and Safety

The International Organization for Standardisation, ISO, is the worldwide federation of national standards bodies. ISO defines *quality* as "the totality of features and characteristics of a product or service that bear on its ability to satisfy stated or implied needs."(ISO 8402). The quality of fish and fish products depends on safe and hygienic practices. Outbreaks of fish-borne illnesses are reduced if appropriate practices are followed when handling, manufacturing, refrigerating and transporting fish and fish products. Ensuring standards of quality and safety are high also minimizes the post-harvest losses."

"The fishing industry must ensure that their fish handling, processing and transportation facilities meet requisite standards. Adequate training of both industry and control authority staff must be provided by support institutions, and channels for feedback from consumers established. Ensuring high standards for quality and safety is good economics, minimizing losses that result from spoilage, damage to trade and from illness among consumers."

Fish processing highly involves very strict controls and measurements in order to ensure that all processing stages have been carried out hygienically. Thus, all fish processing companies are highly recommended to join a certain type of food safety system. One of the certifications that are commonly known is the Hazard Analysis Critical Control Points (HACCP).

Hazard Analysis and Critical Control Points

HACCP is a system which identifies hazards and implements measures for their control. It was first developed in 1960 by NASA to ensure food safety for the manned space programme. The main objectives of NASA were to prevent food safety problems and control food borne diseases. HACCP has been widely used by food industry since the late 1970 and now it is internationally recognized as the best system for ensuring food safety.

"The Hazard Analysis and Critical Control Points (HACCP) system of assuring food safety and quality has now gained worldwide recognition as the most cost-effective and reliable system available. It is based on the identification of risks, minimizing those risks through the design and layout of the physical environment in which high standards of hygiene can be assured, sets measurable standards and establishes monitoring systems. HACCP also establishes procedures for verifying that the system is working effectively. HACCP is a sufficiently flexible system to be successfully applied at all critical stages — from harvesting of fish to reaching the consumer. For such a system to work successfully, all stakeholders must cooperate which entails increasing the national capacity for introducing and maintaining HACCP measures. The system's control authority needs to design and implement the system, ensuring that monitoring and corrective measures are put in place."

HACCP is endorsed by the:

- FAO (Food and Agriculture Organization)
- Codex Alimentarius (a commission of the United Nations)
- FDA (US Food and Drug Administration)
- European Union
- WHO (World Health Organization)

There are seven basic principles:

- Principle 1: Conduct a hazard analysis.
- Principle 2: After assessing all the processing steps, the Critical control point (CCP) is controlled. CCP are points which determine and control significant hazards in a food manufacturing process.
- Principle 3: Set up critical limits in order to ensure that the hazard identified is being controlled effectively.
- Principle 4: Establish a system so as to monitor the CCP.
- Principle 5: Establish corrective actions where the critical limit has not been met. Appropriate actions need to be taken which can be on a short or long-term basis. All records must be sustained accurately.
- Principle 6: Establish authentication procedures so as to confirm if the principles imposed by HACCP documents are being respected effectively and all records are being taken.
- Principle 7: Analyze if the HACCP plan are working effectively.

Final Products

Finfish, or parts of finfish, are typically presented physically for marketing in one of the following forms

- whole fish: the fish as it originally came from the water, with no physical processing

- drawn fish: a whole fish which has been eviscerated, that is, had its internal organs removed
- dressed fish: fish that has been scaled and eviscerated, and is ready to cook.
- pan dressed fish: a dressed fish which has had its head, tail, and fins removed, so it will fit in a pan.
- filleted fish: the "fleshy sides of the fish, cut lengthwise from the fish along the backbone. They are usually boneless, although in some fish small bones called "pins" may be present; skin may be present on one side, too. Butterfly fillets may be available. This refers to two fillets held together by the uncut flesh and skin of the belly"
- fish steaks: large dressed fish can be cut into cross section slices, usually half to one inch thick, and usually with a cross section of the backbone
- fish sticks: "are pieces of fish cut from blocks of frozen fillets into portions at least 3/8-inch thick. Sticks are available in fried form ready to heat or frozen raw, coated with batter and breaded, ready to be cooked"
- fish cakes: are "prepared from flaked fish, potatoes, and seasonings, and shaped into cakes, coated with batter, breaded, and then packaged and frozen, ready-to-be-cooked"
- fish fingers
- fish roe

Value Addition

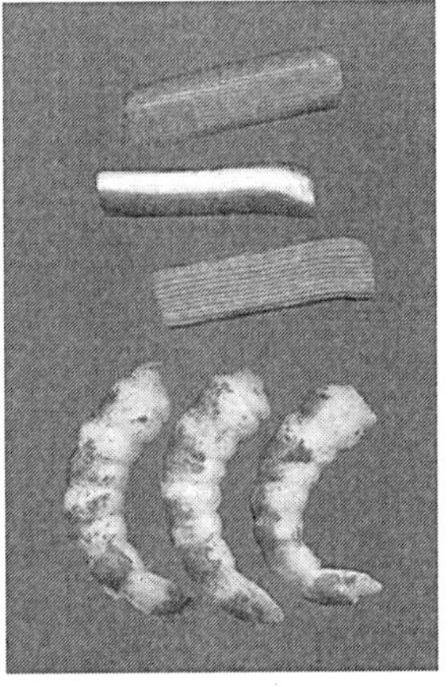

***Figure:** Imitation crab and imitation shrimp made from surimi*

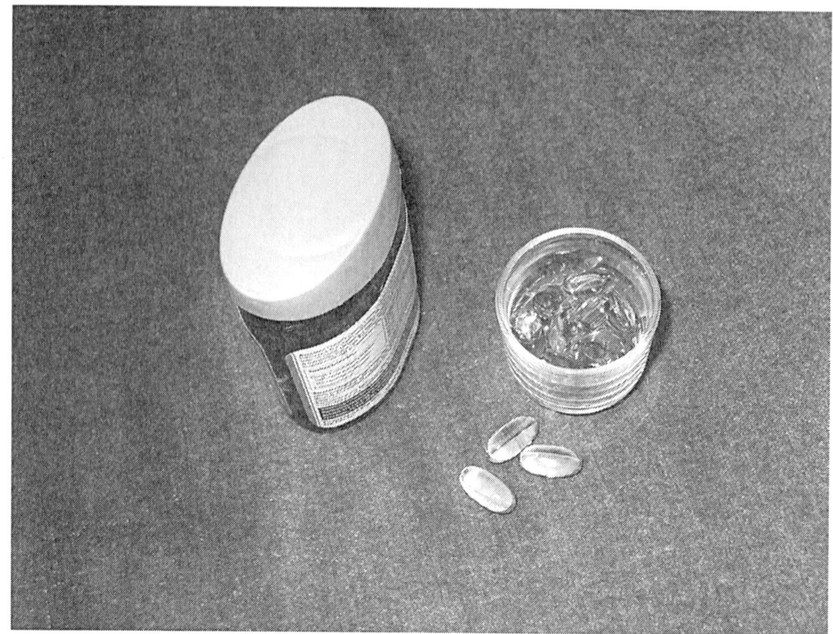

Figure: *Fish oil capsules*

In general value addition means "any additional activity that in one way or the other change the nature of a product thus adding to its value at the time of sale." Value addition is an expanding sector in the food processing industry, especially in export markets. Value is added to fish and fishery products depending on the requirement of different markets. Globally a transition period is taking place where cooked products are replacing traditional raw products in consumer preference.

"In addition to preservation, fish can be industrially processed into a wide array of products to increase their economic value and allow the fishing industry and exporting countries to reap the full benefits of their aquatic resources. In addition, value processes generate further employment and hard currency earnings. This is more important nowadays because of societal changes that have led to the development of outdoor catering, convenience products and food services requiring fish products ready to eat or requiring little preparation before serving."

"However, despite the availability of technology, careful consideration should be given to the economic feasibility aspects, including distribution, marketing, quality assurance and trade barriers, before embarking on a value addition fish process."

- Surimi: Surimi and surimi-based products are an example of value added products. Surimi is prepared from the mechanically deboned, washed (bleached) and stabilised flesh of fish. "It is an intermediate product used in the preparation of a variety of ready to eat seafood such as kamaboko, fish sausage, crab legs and imitation shrimp products. Surimi-based products are gaining more prominence worldwide, because of the emergence of Japanese restaurants and culinary traditions in North America, Europe and elsewhere. Ideally, surimi should be made from low-value, white fish with excellent gelling ability and which are abundant and available year-round. At present, Alaskan pollack accounts for a large proportion of the surimi supply. Other species, such as sardine, mackerel, barracuda, striped mullet have been successfully used for surimi production."
- Fishmeal and fish oil: "A significant proportion of the world catch (20 percent) is processed into fishmeal and fish oil. Fishmeal is a ground solid product that is obtained by removing most of the water and some or all of the oil from fish or fish waste. This industry was launched in the 19th century, based mainly on surplus catches of herring from seasonal coastal fisheries to produce oil for industrial uses in leather tanning and in the production of soap, glycerol and other non-food products. Presently, it uses small oily fish to produce fishmeal and oil. It is worthy to mention that, only where it is uneconomic or impracticable for human consumption, should the catch be reduced to fishmeal and oil. Indeed, cycling fish through poultry or pigs is a loss because there is a need for 3 kg of edible fish to produce approximately 1 kg of edible chicken or pork.

Industrial Rendering

Rendering is a process that converts waste animal tissue into stable, value-added materials. Rendering can refer to any processing of animal products into more useful materials, or more narrowly to the rendering of whole animal fatty tissue into purified fats like lard or tallow. Rendering can be carried out on an industrial, farm, or kitchen scale.

The majority of tissue processed comes from slaughterhouses, but also includes restaurant grease and butcher shop trimmings, expired meat from grocery stores, and the carcasses of euthanized and dead animals from animal shelters, zoos and veterinarians. This material can include the fatty tissue, bones, and offal, as well as entire carcasses of animals condemned at slaughterhouses, and those that have died on farms, in transit, etc. The most common animal sources are beef, pork, sheep, and poultry.

The rendering process simultaneously dries the material and separates the fat from the bone and protein. A rendering process yields a fat commodity (yellow grease, choice white grease, bleachable fancy tallow, etc.) and a protein meal (meat and bone meal, poultry byproduct meal, etc.).

Rendering plants often also handle other materials, such as slaughterhouse blood, feathers and hair, but do so using processes distinct from true rendering.

Process Variations

The rendering process varies from plant to plant in many ways.

1. Whether the end products are to be used as human food is based on the type of raw material and the processing methods.
2. Whether the end products are to be used as animal or pet food.
3. The material may be processed wet or dry. In wet processing, either boiling water or steam is added to the material, causing fat to rise to the surface; in dry processing, fat is released by dehydrating the raw material.
4. The temperature range used, whether high or low.
5. Processing may be either in discrete batches or in a continuous process.
6. The processing plant may be operated by an independent company that collects the material on the open market, or by the packing plant that produced the material.

Rendering Processes for Edible Products

Edible rendering processes are basically meat processing operations and produce lard or edible tallow for use in food products. Edible rendering is generally carried out in a continuous process at low temperature (less than the boiling point of water). The process usually consists of finely chopping the edible fat materials (generally fat trimmings from meat cuts), heating them with or without added steam, and then carrying out two or more stages of centrifugal separation. The first stage separates the liquid water and fat mixture from the solids. The second stage further separates the fat from the water. The solids may be used in food products, pet foods, etc., depending on the original materials. The separated fat may be used in food products, or if in surplus, it may be diverted to soap making operations. Most edible rendering is done by meat packing or processing companies.

One edible product is greaves, which is the unmeltable residue left after animal fat has been rendered.

An alternative process cooks slaughterhouse offal to produce a thick, lumpy "stew" which is then sold to the pet food industry to be used principally as tinned cat and dog foods. Such plants are notable for the offensive odour that they can produce and are often located well away from human habitation.

Rendering Processes for Inedible Products

Materials that for aesthetic or sanitary reasons are not suitable for human food are the feedstocks for inedible rendering processes. Much of the inedible raw material is rendered using the "dry" method. This may be a batch or a continuous process in which the material is heated in a steam-jacketed vessel to drive off the moisture and simultaneously release the fat from the fat cells. The material is first ground, then heated to release the fat and drive off the moisture, percolated to drain off the free fat, and then more fat is pressed out of the solids, which at this stage are called "cracklings" or "dry-rendered tankage". The cracklings are further ground to make meat and bone meal. A variation on a dry process involves finely chopping the material, fluidizing it with hot fat, and then evaporating the mixture in one or more evaporator stages. Some inedible rendering is done using a wet process, which is generally a continuous process similar in some ways to that used for edible materials. The material is heated with added steam and then pressed to remove a water-fat mixture which is then separated into fat, water and fine solids by stages of centrifuging and/or evaporation. The solids from the press are dried and then ground into meat and bone meal. Most independent renderers process only inedible material.

Meat Packing Plant

The meat packing industry handles the slaughtering, processing, packaging, and distribution of animals such as cattle, pigs, sheep and other livestock. The industry is primarily focused on producing meat for human consumption, but it also yields a variety of by-products including hides, feathers, dried blood, and, through the process of rendering, fat such as tallow and protein meals such as meat & bone meal.

In the U.S. and some other countries, the facility where the meat packing is done is called a *meat packing plant*; in New Zealand, where most of the products are exported, it is called a *freezing works*. An abattoir is a place where animals are slaughtered for food. The meat packing industry grew with the construction of the railroads and methods of refrigeration. Railroads made possible the transport of stock to central points for processing, and the transport of products throughout the nation.

Slaughterhouse

A slaughterhouse or *abattoir* or meatworks is a facility where animals are killed for consumption as food products. Slaughterhouses which process meat not intended for human consumption are sometimes referred to as Knacker's yards or Knackeries.

In the United States, around nine billion animals are slaughtered every year. (this includes about 150.4 million cattle, bison, sheep, hogs, and goats and 8.9 billion chickens, turkeys, and ducks; in 2009, 13,450,000 long tons (13,670,000 t) of beef were consumed in the U.S. alone. In Canada, 650 million animals are killed annually. In the European Union, the annual figure is 300 million cattle, sheep, and pigs, and four billion (an unverified number) chickens.

Slaughtering animals on a large scale poses significant logistical problems and public health requirements. Public aversion to meat packing in many cultures influences the location of slaughterhouses. In addition, some religions stipulate certain conditions for the slaughter of animals.

There has been criticism of the methods of transport, preparation, herding, and killing within some slaughterhouses, and in particular of the speed with which the slaughter is sometimes conducted. Investigations by animal welfare and animal rights groups have indicated that in some cases animals are skinned or gutted while alive and conscious. In some cases animals are driven for hundreds of miles to slaughterhouses in conditions that often result in injuries and death en route. Slaughtering animals is opposed by animal rights groups on ethical grounds.

Productivity

Typically 45–50% of the animal can be turned into edible products (meat). About 5-15% is waste, and the remaining 40–45% of the animal is turned into byproducts such as leather, soaps, candles (tallow), and adhesives.

History

Slaughterhouses have existed as long as there have been settlements too large for individuals to rear their own stock for personal digestion and consumption.

Early maps of London show numerous stockyards in the periphery of the city, where slaughter occurred in the open air. A term for such open-air slaughterhouse is a shambles. There are streets named "The Shambles" in some English towns (e.g. Worcester, York) which got their name from having been the site on which butchers killed and prepared animals for consumption .

A "private slaughterhouse" is unregulated and can exist on private property, such as a backyard or butcher's shed. In the eighteenth century, reformers argued that "public slaughterhouses" would be preferable for a number of reasons. Because they would be regulated, they could be monitored and more hygienic, more spacious, and they would remove the sight of animal slaughter from the general public.

Design

In the latter part of the 20th century, the layout and design of most U.S. slaughterhouses was influenced by the work of Dr. Temple Grandin. She suggested that reducing the stress of animals being led to slaughter may help slaughterhouse operators improve efficiency and profit. In particular she applied an understanding of animal psychology to design pens and corrals which funnel a herd of animals arriving at a slaughterhouse into a single file ready for slaughter. Her corrals employ long sweeping curves so that each animal is prevented from seeing what lies ahead and just concentrates on the hind quarters of the animal in front of it. This design – along with the design elements of solid sides, solid crowd gate, and reduced noise at the end point – work together to encourage animals forward in the chute and to not reverse direction.

As of 2011 Grandin claimed to have designed over 54% of the slaughterhouses in the United States as well as many others around the world.

Mobile Design

By 2010 a mobile facility the Modular Harvest System had received USDA approval. It can be moved from ranch to ranch. It consists of three trailers, one for slaughtering, one for consumable body parts and one for other body parts. Preparation of individual cuts is done at a butchery or other meat preparation facility.

Process

The slaughterhouse process differs by species and region and may be controlled by civil law as well as religious laws such as Kosher and Halal laws. A typical U.S. procedure follows:

1. Cattle (mostly steers and heifers, some cows, and even fewer bulls) arrive via truck or rail from a ranch, farm, or feedlot.
2. Place animals in holding pens.
3. Incapacitate them by applying an electric shock of 300 volts and 2 amps to the back of the head, effectively stunning them, or by use of a captive bolt pistol to the front of the cow's head (a pneumatic or cartridge-fired captive bolt). Swine can be rendered unconscious by CO_2/inert gas stunning. (This step is prohibited under strict application of Halal and Kashrut codes.)
4. Hang them upside down by both of their hind legs and place them on the processing line.
5. Sever the carotid artery and jugular vein with a knife. The blood drains from the body, causing death through exsanguination.
6. Remove the head and feet.
7. Cut around the digestive tract to prevent fecal contamination later in the process.
8. Remove the hide/skin by "down pullers", "side pullers" and "fisting" off the pelt (sheep and goats). Hides can also be removed by laying the carcase on a cradle and skinning with a knife.
9. Remove internal organs and inspect them for parasites and signs of disease. Separate the viscera from the heart and lungs, referred to as the "pluck" for inspection. Also separate livers for inspection. Drop or remove tongues from the head, and send the head down the line on head hooks or head racks for inspection of the lymph nodes for signs of systemic disease.
10. A government inspector inspects the carcase for safety. (This inspection is performed by the Food Safety Inspection Service in the U.S., and Canadian Food Inspection Agency in Canada.)
11. Reduce levels of bacteria using interventions such as steam, hot water, and organic acids.
12. Optionally electrically stimulate cattle and sheep (only) to improve meat tenderness.
13. Chill carcases to prevent the growth of microorganisms and to reduce meat deterioration while the meat awaits distribution.
14. Cut the chilled carcase into primal cuts, subprimals and/or leave intact as a "side" of meat. Beef and horse carcases are always split in half and then quartered, pork is split into sides only and goat/veal/mutton and lamb is left whole.

15. The remaining carcase may be further processed to extract any residual traces of meat, usually termed advanced meat recovery or mechanically separated meat, for human or animal consumption.
16. Materials such as bone, lard or tallow, are sent to a rendering plant. Also, lard and tallow can be used for the production of biodiesel or heating oil.
17. The wastewater, consisting of blood and fecal matter, generated by the slaughtering process is sent to a waste water treatment plant.
18. The meat is transported to distribution centres that then distribute to retail markets.

International Variations

The standards and regulations governing slaughterhouses vary considerably around the world. In many countries the slaughter of animals is regulated by custom and tradition rather than by law. In the non-Western world, including the Arab world, the Indian sub-continent, etc., both forms of meat are available: one which is produced in modern mechanized slaughterhouses, and the other from local butcher shops.

In some communities animal slaughter may be controlled by religious laws, most notably halal for Muslims and kashrut for Jewish communities. These both require that the animals being slaughtered should be conscious at the point of death, and as such animals cannot be stunned prior to killing. This can cause conflicts with national regulations when a slaughterhouse adhering to the rules of religious preparation is located in some Western countries. In Islamic and Jewish law, captive bolts and other methods of pre-slaughter paralysis are generally not permissible, due to it being forbidden for an animal to be killed prior to slaughter. Various halal food authorities have more recently permitted the use of a recently developed fail-safe system of head-only stunning where the shock is less painful and non-fatal, and where it is possible to reverse the procedure and revive the animal after the shock.

In many societies, traditional cultural and religious aversion to slaughter led to prejudice against the people involved. In Japan, where the ban on slaughter of livestock for food was lifted only in the late 19th century, the newly found slaughter industry drew workers primarily from villages of *burakumin*, who traditionally worked in occupations relating to death (such as executioners and undertakers). In some parts of western Japan, prejudice faced by current and former residents of such areas (*burakumin* "hamlet people") is still a sensitive issue. Because of this, even the Japanese word for "slaughter" (`\ºk *tosatsu*) is deemed politically incorrect by some pressure groups as its inclusion of the kanji for "kill" (ºk) supposedly portrays those who practice it in a negative manner.

Some countries have laws that exclude specific animal species or grades of animal from being slaughtered for human consumption, especially those that are taboo food. The former Indian Prime Minister Atal Bihari Vajpayee suggested in 2004 introducing legislation banning the slaughter of cows throughout India, as Hinduism holds cows as sacred and considers their slaughter unthinkable and offensive. This was often opposed on grounds of religious freedom. The slaughter of cows and the importation of beef into the nation of Nepal are strictly forbidden.

Law

Most countries have laws in regard to the treatment of animals at slaughterhouses. In the United States, there is the Humane Slaughter Act of 1958, a law requiring that all swine, sheep, cattle, and horses be stunned unconscious with application of a stunning device by a trained person before being hoisted up on the line. There is some debate over the enforcement of this act. This act, like those in many countries, exempts slaughter in accordance to religious law, such as kosher shechita and dhabiha halal. Most strict interpretations of kashrut require that the animal be fully sensible when its carotid artery is cut.

The novel *The Jungle* detailed unsanitary conditions in slaughterhouses and the meatpacking industry during the 1800s. This led directly to an investigation commissioned directly by President Theodore Roosevelt, and to the passage of the Meat Inspection Act and the Pure Food and Drug Act of 1906, which established the Food and Drug Administration. A much larger body of regulation deals with the public health and worker safety regulation and inspection.

Animal Welfare Concerns

For her book *Slaughterhouse*, Gail Eisnitz, chief investigator for the Humane Farming Association (HFA), interviewed slaughterhouse workers in the U.S. who say that, because of the speed with which they are required to work, animals are routinely skinned while apparently alive, and still blinking, kicking, and shrieking. Eisnitz argues that this is not only cruel to the animals, but also dangerous for the human workers, as cows weighing several thousands of pounds thrashing around in pain are likely to kick out and debilitate anyone working near them.

According to the HFA, Eiznitz interviewed slaughterhouse workers representing over two million hours of experience, who, without exception, told her that they have beaten, strangled, boiled, and dismembered animals alive, or have failed to report those who do. The workers described the effects the violence has had on their personal lives, with several admitting to being physically abusive or taking to alcohol and other drugs.

The HFA alleges that workers are required to kill up to 1,100 hogs an hour, and end up taking their frustration out on the animals. Eisnitz interviewed one worker, who had worked in ten slaughterhouses, about pig production. He told her:

"Hogs get stressed out pretty easy. If you prod them too much, they have heart attacks. If you get a hog in the chute that's had the shit prodded out of him and has a heart attack or refuses to move, you take a meat hook and hook it into his bunghole. You try to do this by clipping the hipbone. Then you drag him backwards. You're dragging these hogs alive, and a lot of times the meat hook rips out of the bunghole. I've seen hams — thighs — completely ripped open. I've also seen intestines come out. If the hog collapses near the front of the chute, you shove the meat hook into his cheek and drag him forward. "

Over the last few decades, some research has been done toward making slaughterhouses more humane; one well-known scientist in this field is Temple Grandin.

Fish

Historically, some doubted that fish could experience pain. However, laboratory experiments have shown that fish do react to painful stimuli (e.g. injections of bee venom) in a similar way to mammals. The expansion of fish farming as well as animal welfare concerns in society has led to research into more humane and faster ways of killing fish. In large-scale operations like fish farms, stunning fish with electricity or putting them into water saturated with nitrogen so that they cannot breathe, results in death more rapidly than just taking them out of the water. For sport fishing, it is recommended that fish be killed soon after catching them by hitting them on the head followed by bleeding out, or by stabbing the brain with a sharp object (called pithing or ike jime in Japanese).

Major Slaughterhouses

The largest slaughterhouse in the world is operated by the Smithfield Packing Company in Tar Heel, North Carolina. It is capable of butchering over 32,000 pigs a day. In the US, the majority of major meat packing plants are located in the Midwestern and High Plains regions.

Sugar Industry

A sugar refinery is a refinery which processes raw sugar into white refined sugar.

Many cane sugar mills produce raw sugar, which is sugar that still contains molasses, giving it more colour (and any associated nutrients) than the white sugar which is normally consumed in households and used as an ingredient in soft drinks and foods. While cane sugar does not strictly need refining, beet sugar is almost always refined to remove the strong, almost always unwanted, taste of beets from it thus also removing nutrients that are found in beets. The refined sugar produced is more than 99 percent pure sucrose. Whereas many sugar mills only operate during a limited time of the year during the cane harvesting period, many sugar refineries work the whole year round.

Raw sugar is either processed into white refined sugar in local refineries, and sold to the local industry and consumers, or it is exported and refined in the country of destination. Sugar refineries are often located in heavy sugar-consuming regions such as North America, Europe, and Japan. Since the 1990s many state-of-the art sugar refineries have been built in the MENA region, e.g. in Dubai, Saudi Arabia and Algeria. The world's largest sugar refinery company is American Sugar Refining with facilities in North America and Europe.

Affination

Figure: *Raw sugar storage in a sugar refinery*

The raw sugar is stored in large warehouses and then transported into the sugar refinery by means of transport belts. In the traditional refining process, the raw sugar is first mixed with heavy syrup and centrifuged to wash away the outer coating of the raw sugar crystals, which is less pure than the crystal interior. Many sugar refineries today buy high pol sugar and can do without the affination process.

Purification

The remaining sugar is then dissolved to make a syrup (about 70 percent by weight solids), which is clarified by the addition of phosphoric acid and calcium hydroxide that combine to precipitate calcium phosphate. The calcium phosphate particles entrap some impurities and absorb others, and then float to the top of the tank, where they are skimmed off.

After any remaining solids are filtered out, the clarified syrup is decolourized by filtration through a bed of activated carbon or, in more modern plants, ion-exchange resin.

Sugar House

Figure: *Vacuum pans*

The purified syrup is then concentrated to supersaturation and repeatedly crystallized under vacuum to produce white refined sugar. As in a sugar mill, the sugar crystals are separated from the mother liquor by centrifuging. To produce granulated sugar, in which the individual sugar grains do not clump together, sugar must be dried.

Sugar Drying and Storage

Drying is accomplished first by drying the sugar in a hot rotary dryer, and then by blowing cool air through it for several days in so-called conditioning silos. The finished product is stored in large concrete or steel silos. It is shipped in bulk, big bags or 25 – 50 kg bags to industrial customers or packed in consumer-size packages to retailers.

The dried sugar must be handled with caution, as sugar dust explosions are possible. A sugar dust explosion which led to 13 fatalities was the 2008 Georgia sugar refinery explosion in Port Wentworth, GA.

Molasses

Factory Automation in Sugar Refineries

As in many other industries factory automation has been promoted heavily in sugar refineries in recent decades. The production process is generally controlled by a central process control system, which directly controls most of the machines and components. Only for certain special machines such as the centrifuges in the sugar house decentralized PLCs are used for security reasons.

8

Food Engineering

Food engineering is a multidisciplinary field of applied physical sciences which combines science, microbiology, and engineering education for food and related industries. Food engineering includes, but is not limited to, the application of agricultural engineering, mechanical engineering and chemical engineering principles to food materials. Food engineers provide the technological knowledge transfer essential to the cost-effective production and commercialization of food products and services.

Food engineering is a very wide field of activities. Prospective major employers for food engineers include companies involved in food processing, food machinery, packaging, ingredient manufacturing, instrumentation, and control. Firms that design and build food processing plants, consulting firms, government agencies, pharmaceutical companies, and health-care firms also hire food engineers. Among its domain of knowledge and action are:

- research and development of new foods, biological and pharmaceutical products
- development and operation of manufacturing, packaging and distributing systems for drug/food products
- design and installation of food/biological/pharmaceutical production processes
- design and operation of environmentally responsible waste treatment systems
- marketing and technical support for manufacturing plants.

Topics in food engineering In the development of food engineering, one of the many challenges is to employ modern tools and knowledge, such as computational materials science and nanotechnology, to develop new products and processes. Simultaneously, improving quality, safety, and security remain critical issues in food engineering study. New packaging materials and techniques are being developed to provide more protection to foods, and novel preservation technology is emerging. Additionally, process control and automation regularly appear among the top priorities identified in food engineering. Advanced monitoring and control systems are developed to facilitate automation and flexible food manufacturing. Furthermore, energy saving and minimization of environmental problems continue to be important food engineering issues, and significant progress is being made in waste management, efficient utilization of energy, and reduction of effluents and emissions in food production.

Typical topics include:

- Advances in classical unit operations in engineering applied to food manufacturing
- Progresses in the transport and storage of liquid and solid foods
- Developments in heating, chilling and freezing of foods
- Advanced mass transfer in foods
- New chemical and biochemical aspects of food engineering and the use of kinetic analysis
- New techniques in dehydration, thermal processing, non-thermal processing, extrusion, liquid food concentration, membrane processes and applications of membranes in food processing
- Shelf-life, electronic indicators in inventory management, and sustainable technologies in food processing
- Modern packaging, cleaning, and sanitation technologies.

Wholesale and Distribution

A vast global transportation network is required by the food industry in order to connect its numerous parts. These include suppliers, manufacturers, warehousing, retailers and the end consumers. There are also companies that add vitamins, minerals, and other required necessities during processing to make up for those lost during preparation. Wholesale markets for fresh food products have tended to decline in importance in OECD countries. This also occurred in Latin America and some Asian countries as a result of the growth of supermarkets, which procure directly from farmers or through preferred suppliers, rather than going through markets.

***Figure:** Wholesale and Distribution*

The constant and uninterrupted flow of product from distribution centres to store locations is a critical link in food industry operations. Distribution centres run more efficiently, throughput can be increased, costs can be lowered, and manpower better utilized if the proper steps are taken when setting up a material handling system in a warehouse.

Retail

With populations around the world concentrating in urban areas, food buying is increasingly removed from all aspects of food production. This is a relatively recent development, having taken place mainly over the last 50 years. The supermarket is the defining retail element of the food industry. There, tens of thousands of products are gathered in one location, in continuous, year-round supply. Restaurants, cafes, bakeries and mobile trucks also provide opportunities for consumers to purchase food.

Food preparation is another area where the change in recent decades has been dramatic. Today, two food industry sectors are in apparent competition for the retail food dollar. The grocery industry sells fresh and largely raw products for consumers to use as ingredients in home cooking. The food service industry by contrast offers prepared food, either as finished products, or as partially prepared components for final "assembly".

Food Industry Technologies

Modern food production is defined by sophisticated technologies. These include many areas. Agricultural machinery, originally led by the tractor, has practically eliminated human labour in many areas of production. Biotechnology is driving much change, in areas as diverse as agrochemicals, plant breeding and food processing. Many other types of technology are also involved, to the point where it is hard to find an area that does not have a direct impact on the food industry. Computer technology is also a central force, with computer networks and specialized software providing the support infrastructure to allow global movement of the myriad components involved.

Marketing

As consumers grow increasingly removed from food production, the role of product creation, advertising, and publicity become the primary vehicles for information about food. With processed food as the dominant category, marketers have almost infinite possibilities in product creation.

Media & Marketing

A key tool for FMCG marketing managers targeting the supermarket industry includes national titles like *The Grocer* in the U.K., *Checkout* in Ireland, *Progressive Grocer* in the U.S., and *Private Label Europe* for the entire of the European Union.

Labour and education

Until the last 100 years, agriculture was labour intensive. Farming was a common occupation and millions of people were involved in food production. Farmers, largely trained from generation to generation, carried on the family business. That situation has changed dramatically today. In North America, only a few decades ago over 50% of the population were farm families. Now, that figure is around 1-2%, and about 80% of the population lives in cities. The food industry as a complex whole requires an incredibly wide range of skills. Several hundred occupation types exist within the food industry.

Quality Control/quality Assurance and International Trade

The international trade in processed fruits and vegetables is very large with an ever increasing number of different types being processed and exported. Whereas once, processing was limited to mostly temperate climate fruits and vegetables, the change has now broadened to include tropical and subtropical types.

The reasons are twofold. Firstly, consumers' dietary habits have become more diverse so that, for example people living in North America may very well like fruit and vegetables grown in Africa or Asia. Secondly, processing techniques, whether they be for canning, freezing or drying, have been improved to an extent where final product is palatable, nutritious and of long and reliable shelf life.

Many developing countries have taken advantage of the continuing worldwide demand for processed fruits and vegetables and earned valuable foreign exchange from exports of products to profitable markets.

The export quality control and inspection of processed fruits and vegetables is directed at ensuring that the final products:

have been processed in a registered export establishment that is constructed, equipped and operated in an hygienic and efficient manner;

conform to the requirements of the export regulations for processed fruits and vegetables, and those of the importing country, in respect of such things as quality grades, defects, ingredients, packaging materials, styles, additives, contaminants, fill of container, drained weight; and, conform to labelling requirements.

Inspection and Certification Procedures

In most countries, in processing fruits and vegetables for export, it is not customary to apply continuous inspection as it is in the case of meat. Few, if any, importing countries require it, and the nature of the products themselves is such that only part time check inspection is required during processing together with statistically based inspection, including sampling and analysis, of final product.

However, in circumstances where an establishment is processing export product for the first time, it can be argued that there is an advantage in adopting continuous inspection until the operation is satisfactorily established.

In any event, inspection of raw materials should be carried out at the commencement of each processing run to ensure that only sound fruit or vegetables of sufficient maturity (degree of ripeness) are used for processing. Sampling checks of raw materials should be carried out as frequently as the inspector thinks necessary.

The inspector must ensure that adequate hygiene practices are followed during the processing of the product. For example, in the case of canned and frozen products and other processing methods, raw materials should be washed absolutely clean so that fruit and vegetables entering the processing line are free from dirt, superficial residues of agricultural chemicals, insects and extraneous plant material.

In the case of dried product, especially where the raw material is sun-dried on drying greens or racks, care must be taken to minimize contamination by bird and animal droppings, dust and extraneous plant material. It is often necessary to wash the dried product to ensure cleanliness of the final product.

In the case of canning and freezing, the inspector must obtain full details of the processing programme for at least the following day from management, so that an adequate inspection programme can be scheduled.

In much the same way as for fresh fruit and vegetables, the inspector must also be aware of the pesticides and other chemicals used in the production of the raw materials. Necessary laboratory analyses can then be arranged to ensure residue levels in the final product do not exceed tolerances adopted by importing countries.

At the commencement of and during processing, the inspector should pay attention to the state of raw materials, the preparation of raw materials for processing (peeling, slicing, dicing, blanching, etc.), preparation and density of packing medium (sugar syrup, salt brine, etc.), the state of cans or containers to be used (cleanliness and strength), the cooking or freezing process (time/temperature relationship), can filling and closure and can/container storage.

After processing, the inspector should check the final product to ensure the drained or thawed weight, the vacuum and headspace, packing medium strength and that can/container conditions are satisfactory. Statistically based sampling plans should be adopted for the examination of final product to ensure it meets the requirements of the export regulations.

The labelling applied to cans/containers should also be checked to ensure both their correctness and compliance with the export regulations and the requirements of those countries in which the product is to be marketed.

Cans should also be examined to make sure that the correct embossing relating to the product, its date of production and the registered number of the export establishment has been applied.

Each establishment registered for the export of processed fruit and vegetables or for canned or frozen foods should have its own quality laboratory sufficiently equipped and staffed to carry out physical, chemical and microbiological examinations of the goods.

Inspectors should have access to the laboratory facilities and the establishment's quality control records as and when required. Independent laboratory examination of product should be made by the agency having responsibility for export on the basis of a statistically developed sampling plan.

In those countries where fruit and vegetable production is a seasonal event, processing for export generally takes place at the time of peak production and then declines, often to a halt, as the supply of raw materials declines. As a result, most export establishments produce at their peak of production far more product than they export at that time.

Therefore, most manufacturers find it necessary to store product for considerable periods before it is exported. Thus, proper storage is essential if the product is to retain its quality and cans remain untarnished. Inspectors should regularly inspect storage facilities, noting their conditions and that of the stored product, looking for signs of deterioration such as pest infestation and rusting of cans.

Prior to export, the exporter should be required to notify the export quality and inspection agency of his intention to export in accordance with the provisions of the export processed fruits and vegetables regulations and on the prescribed "Notice of Intention to Export" form.

The notice should be submitted in sufficient time before the shipment date to enable the product to be inspected satisfactorily; the intensity of inspection depending on the original state of the product, the conditions under which it has been stored and the length of storage. When product is approved, the agency will issue the exporter an "Export Permit" authorizing Custom's clearance of the product.

Labelling

Customers and consumers expect the labelling on food to be a true description of what they are buying.

Misleading or fraudulent labelling is an unfair trade practice that cannot be tolerated. Most countries now have labelling laws stipulating how foods are to be labelled and what information labels must contain. Most, if not all of those laws have in common requirement that the label should bear:

- a statement of identity and a true, as distinct from misleading, description of the product;
- a declaration of net contents (weight or number of pieces);
- the name and address of the manufacturer, packer, distributor or consignee, and
- a list of ingredients (in descending order of volume or weight).

In addition, labels may also be required to include, amongst other things, the country of origin, date of manufacture or packing, a use-by or expiry date, nutritional qualities or values of the food, storage directions, a quality grade and directions for preparing the food.

More frequently than is often realized, consignments of food exports arriving on foreign markets are not permitted entry because the labelling does not comply with the mandatory requirements of the importing country.

This sometimes results in consignments being rejected, but more often in them being withheld from entry until the labelling is corrected or new labelling applied. In either case, trade is interrupted and the cost involved may make sales unprofitable. It is essential therefore, that exporters be familiar with the food labelling requirements of importing countries.

Export Quality Control and Inspection Systems for Foods

With the advent and development of a food consciousness amongst consumers, stimulated by the work of the Joint FAO/WHO Codex Alimentarius Commission through its elaboration of food standards, codes of hygienic practice and the Code of Ethics for International Trade in Food, an increasing number of countries have adopted sophisticated food laws and established food control agencies, some with the aid of FAO.

Consequently, those countries no longer accept products on trust that they are satisfactory, but instead, demand that food imports meet the requirements of their food laws and pass inspection by their control agencies. Moreover, many of them require exporting countries to certify that products comply with their national legislation and some also require additional special declarations.

As a result of these developments the emphasis of activity of Export Quality Control and Inspection Systems has changed. Although most of them still establish their own standards of quality control and adopt standards for foods for export, most of their effort and resources are now directed at ensuring that foods for export meet the mandatory requirements of importing countries and providing the necessary associated certification. To do otherwise is to invite either the detention or, at worst, rejection of product at point of entry.

Detentions and Rejections

Food exporting countries can no longer assume that there is a good chance that products not complying with the requirements of importing countries will escape the inspection at the point of entry.

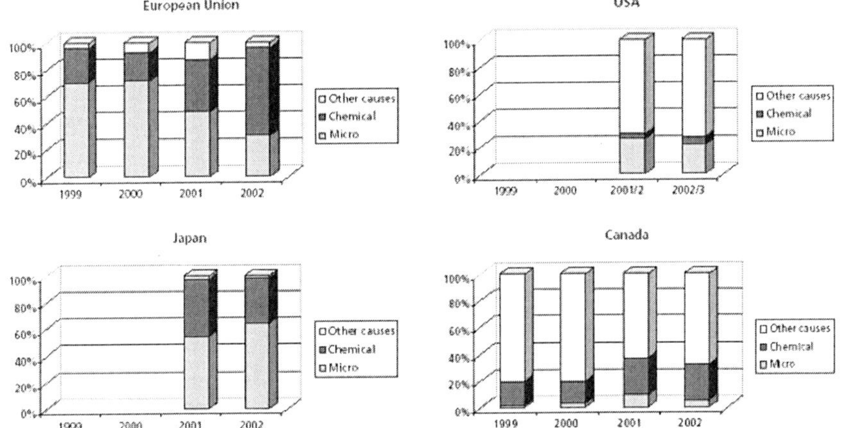

Figure: *Discusses in more detail the disparity between the data recorded for the causes of border cases, and the difficulty this proses for comparison between importing regions.*

Details of foods imports released by the United States Food and Drug Administration (FDA) indicate that significant quantities of product are at least detained, and at worst rejected, because they fail to meet U.S. food laws.

Reasons given for the detentions include:

- non compliance with labelling requirements;
- decomposition;
- insect and animal filth and damage;
- use of prohibited additives;
- non compliance with requirements of the U.S. low acid canned food regulations;
- heavy metal contamination;

- excessive levels of pesticide residues;
- excessive levels of mycotoxin;
- mould infestation;
- microbiological contamination;
- swollen and otherwise faulty cans.

The message for food exporting countries is quite clear - ensure your products comply with the mandatory requirements of importing countries or run the very real risk of having them rejected at considerable financial loss to the exporter and the country and resulting in damage to the commercial reputation of both.

While the foregoing relates to the U.S.A. experience, because it is the only country that currently publishes data about detentions and rejections of food imports it can be assumed that record more or less reflects the experience of other food importing countries. It might well be asked why such significantly high levels of detentions and rejections of food imports take place.

Undoubtedly the reasons are many and varied. However, the evidence shows that the most important reasons include:

- the inability of some export food industries, especially in developing countries, to handle, process, package and transport products to meet the mandatory requirements of importing countries;
- lack of awareness by food exporting countries of the mandatory requirements of importing countries, including certification;
- lack of adequate export control programmes and related agencies in food exporting countries, preventing them from exercising the necessary product surveillance and giving reliable and credible certification, and
- a lack of communication, between food control authorities and agencies in exporting and importing countries.

All four can be remedied by governments if they possess sufficient political will and take the necessary steps to do so.

Good Manufacturing Practices (gmp); Hygiene Requirements

Disease Control: Any person who has an illness, open lesions, including boils, sores, infected wounds, or any other abnormal source of microbial contamination must not work in any operation (in a food processing centre) which could result in the food, food-contact surface, or food packaging materials becoming contaminated.

Cleanliness

The following applies to people who work in direct contact with food preparation, food ingredients or surfaces of equipment or utensils that will contact food: they must wear clean outer garments, maintain a high degree of personal cleanliness and conform to hygienic practices while on duty; they must wash their hands thoroughly and, if they are working at a job where it is necessary, they must also sanitize their hands before starting work, after each absence from the workstation and at any other time when the hands have become soiled or contaminated; they must also remove all unsecured jewelry. People who are actually handling food, should remove any jewelry that cannot be properly sanitized from their hands; it is necessary to wear effective hair restraints, such as hairnets, caps, headbands or beard covers; operators must not store clothing or other personal belongings in food processing areas. Also, eating food, drinking beverage or using tobacco (in any form) must not be allowed in food processing area; all necessary steps have to be taken by supervisors to prevent operators from contaminating foods with microorganisms or foreign substances such as perspiration, hair, cosmetics, tobacco, chemicals and medicants.

Education and Training

Persons who are monitoring the sanitation programmes must have the education and/or experience to demonstrate that they are qualified. Food handlers and supervisors should receive training that will make them aware of the danger of poor personal hygiene and unsanitary work habits.

***Figure:** Education and Training Center*

Supervision

Someone must be assigned the responsibility that all personnel will comply with all the requirements of these GMP's.

Plants and Grounds

Grounds around a food processing centre which are under the control of this centre must be free from conditions such as: improperly stored equipment; litter, waste or refuse; uncut weeds or grass close to buildings; excessively dusty roads, yards or parking lots; inadequately drained areas - potential foot-borne filth or breeding places for insects or microorganisms; inadequately operated systems for waste treatment and disposal.

Plant construction and design shall: provide enough space for sanitary arrangement of equipment and storage of materials; floors, walls and ceilings must be constructed so that they are cleanable and must be kept clean and in good repair; separate by partition, location, time and other means, any operations that may cause cross-contamination of food products with undesirable microorganisms, chemicals, filth or other extraneous material; provide effective screening or other protection to keep out birds, animals and vermin such as insects and rodents. provide adequate ventilation to prevent contamination of foods with odours, noxious fumes or vapours (including steam); light bulbs, skylights or any other glass must be of the safety type or protected so that glass contamination cannot occur in case of breakage.

Sanitary Operations

General Maintenance: The plant and all fixtures must be kept in good repair and be maintained in a sanitary condition. Cleaning operations must be conducted in a manner that will minimize the possibility of contaminating foods or equipment surfaces that contact food.

Pest control:

- No animals or birds are allowed anywhere in the plant
- Programmes must be in effect to prevent contamination by animals, birds and pests, such as rodents and insects;
- Insecticides and rodenticides may be used as long as they are used properly (according to label instructions);
- These pesticides must not contaminate food or packaging materials with illegal residues;

Sanitation of equipment and utensils:

- Utensils and equipment surfaces that are in contact with food must be cleaned as often as necessary to prevent food contamination;
- Equipment surfaces that are not in contact with food should be cleaned as frequently as necessary to minimize accumulation of dust, dirt, food particles, etc.
- Single-service articles such as disposable utensils, paper cups, paper towels, etc., should be:
- Stored in appropriate containers;
- Handled, dispensed, used and disposed of in a manner that prevents contamination of food or equipment;
- Where there is the possibility of introducing undesirable microorganisms into food, all utensils and equipment surfaces that contact food must be cleaned and sanitized before use and following any interruption during which they may have become contaminated;
- When utensils or equipment are used in a continuous production operation, they must be cleaned and sanitized on a predetermined schedule;
- Any facility, procedure, machine or device may be used for cleaning and sanitizing, as long as it has been established that the procedure will do the job effectively.

10.2.3.4. Storage and handling of clean portable equipment and utensils

a) This refers to portable equipment or utensils which have surfaces that will contact foods;

b) When such equipment or utensils have been cleaned and sanitized, they should be stored in a manner that will protect the food contact surfaces from splash, dust and other contamination.

Sanitary Facilities and Vontrols

Water Supply: Any water that comes into contact food or processing equipment must be safe and of adequate sanitary quality.

Sewage Disposal: Must flow into an adequate sewage system or disposed of through other adequate means.

Plumbing: Must be of adequate size and design to:

a. Supply enough water to areas in the plant where it is needed;
b. Properly convey sewage or disposable liquid waste from the plant;
c. Not create a source of contamination or unsanitary condition;
d. Provide adequate floor drainage where hosing-type cleaning is done or where operations discharge water or liquid waste onto the floor;
e. Insure that there is no backflow from cross-connection between piping systems that discharge waste water or sewage, and those that carry water for food or food manufacturing.

Toilet facilities

a. Toilets and hand-washing facilities must be provided inside the fruit and vegetable processing centres;
b. Toilet tissue must be provided;
c. Toilets must be kept sanitary and in good repair;
d. Toilet rooms must have self-closing doors;
e. Toilet rooms must not open directly into areas where food is exposed unless steps have been taken to prevent airborne contamination (example: double doors, positive airflow, etc.);
f. Signs must be posted that direct employees to wash their hands with soap or detergent after using the toilet.

Hand-washing Facilities

a) Adequate and convenient hand-washing and, if necessary, hand-sanitizing facilities must be provided anywhere in the plant where the nature of employees jobs requires that they wash, sanitize and dry their hands;
b) These hand-washing facilities must provide:
   - Running water at a suitable temperature;
   - Effective hand-cleaning and hand-sanitizing preparations;
   - Clean towel service or suitable drying devices;
   - Easily cleanable waste receptacle;
   - Water control valves designed and constructed to protect against recontamination of clean, sanitized hands;
   - Signs directing employees handling unprotected food to wash and, if appropriate, sanitize theirs hands before starting work, after each absence from the workstation, and any other time when the hands have become soiled or contaminated.

Rubbish and offal disposal must be handled in such a manner that they do not serve to attract or harbour pests or create contaminating conditions.

Equipment and Utensils

Figure: *About Kitchen And Recipe*

a. Equipment and utensils must be designed and constructed so that they are adequately cleanable and will not adulterate food with lubricants, fuel, metal fragments, contaminated water, etc.

b. Equipment should be installed so that it, and the area around it, can be cleaned;

c. Food contact surfaces shall be made of nontoxic materials and must be corrosion-resistant;

d. Seams on food contact surfaces shall be smoothly bonded, or maintained in order to minimize the accumulation of food particles, dirt and organic matter;

e. Equipment in processing areas that does not come into contact with food shall be constructed so that it can be kept clean;

f. Holding, conveying and manufacturing systems, including gravimetric, pneumatic, closed and automated systems, shall be maintained in a sanitary condition;

g. Each freezer and cold storage compartment shall have an indicating thermometer, temperature measuring or recording device, and should have an automatic control for regulating temperature, or an automatic alarm system to indicate a significant temperature change;

h. Instruments and controls used for measuring, regulating or recording temperatures, pH, acidity, water activity, etc. shall be adequate in number, accurate and maintained.

Processes and Controls

There must be an individual who is responsible for supervising the overall sanitation of the plant.

Raw Materials and Ingredients

a. Must be inspected and sorted to insure that they are clean, wholesome and fit for processing into human food;

b. Must be stored under conditions that will protect against contamination and minimize deterioration;

c. Must be washed or cleaned to remove soil and other contamination:

   - Water used for washing, rising or conveying food products must be of sanitary quality;
   - Water must not be reused for washing, rinsing or conveying if contamination of food may result;
   - Containers and carriers (such as trucks or railcars) should be inspected to assure that their condition has not contaminated raw ingredients;

d. Raw materials shall not contain levels of microorganisms that may produce food poisoning or other disease, or they shall be pasteurized or otherwise treated during manufacturing operations so that the product will not be adulterated;

e. Materials susceptible to contamination with natural toxins, e.g., aflatoxin, shall comply with national and international official levels before they are incorporated into the finished food;

f. Materials susceptible to contamination with pests, undesirable microorganisms, or extraneous material, shall comply with national and international regulations, guidelines and defect action levels;

g. Materials shall be stored in containers, and under conditions which protect against contamination;

h. Frozen materials shall be kept frozen. If thawing is required prior to use, it shall be done in a manner that prevents contamination.

Manufacturing Operations

a. Food processing equipment must be keep in a sanitary condition through frequent cleaning and, when necessary, sanitizing. If necessary, such equipment must be taken apart for thorough cleaning.

b. It is necessary to process, package and store food under conditions that will minimize the potential for undesirable microbiological growth, toxin formation, deterioration or contamination. To accomplish this may require careful monitoring of such factors as time, temperature, humidity, pressure, flow rate, etc. The object is to assure that mechanical breakdowns, time delays, temperature fluctuations or other factors do not allow the foods to decompose or become contaminated.

c. Food shall be held under conditions that prevent the growth of undesirable microorganisms as follows:
   - Refrigerated foods shall be maintained at 45° F or below;
   - Frozen foods shall be maintained in a frozen state;
   - Acid or acidified foods to be held in hermetically sealed containers at ambient temperatures shall be heat-treated to destroy mesophyllic microorganisms;

d. Measures such as sterilizing, irradiating, pasteurizing, etc., shall be adequate to destroy or prevent the growth of undesirable microorganisms;

e. Work-in-process shall be protected against contamination;

f. Finished food shall be protected from contamination;

g. Equipment, containers and utensils shall be constructed, handled and maintained to protect against contamination;

h. Measures, e.g., sieves, traps, metal detectors, shall be used to protect against the inclusion of metal or other extraneous material in food;

i. Food or materials that are adulterated shall be disposed of in a manner that prevents other food from being contaminated;

j. Mechanical manufacturing steps such as washing, peeling, etc., shall be performed to protect against contamination by providing adequate protection from contaminants that may drip, drain or be drawn into the food, by adequately cleaning and sanitizing all food-contact surfaces and by using time and temperature controls at and between each manufacturing step;

k. Heat-blanching should be done by heating the food to the required temperature, holding it at this temperature for the required time, and then either rapidly cooling the food or passing it to the next manufacturing step without delay;

l. Filling, assembling, packaging, and other operations shall be performed in such a way that the food is protected against contamination by:
   - Use of a quality control operation in which the Critical Control Points are identified and controlled during manufacturing;
   - Adequate cleaning and sanitizing of all food-contact surfaces and food containers;
   - Using materials for food containers and food-packaging materials that are safe and suitable;
   - Providing physical protection from contamination, particularly airborne contamination;
   - Using sanitary handling procedures.

m. Food such as, but not limited to, dry mixes, nuts, intermediate moisture food, and dehydrated food, that relies on the control of aw for preventing the growth of undesirable microorganisms shall be processed to and maintained at a safe moisture level by:
   - Monitoring the aw of food;
   - Controlling the soluble solids / water ratio in finished food;
   - Protecting finished food from moisture pickup, by use of a moisture barrier, or by other means, so that the Aw of the food does not increase to an unsafe level;

n. Food such as, but not limited to, acid and acidified food, that relies principally on the control of pH for preventing the growth of undesirable microorganisms shall be monitored and maintained at a pH of 4.6 or below by: - Monitoring the pH of raw materials, food in process, and finished food; - Controlling the amount of acid or acidified food added to low-acid food;
      – If ice is used and comes in contact with food products, it must be made from potable water and be in a sanitary condition;

p. Areas and equipment that are used to process human food should not be used to process non-human food-grade animal feed, or inedible products unless there is no possibility of contaminating the human food;

q. A coding system should be utilized that will allow positive lot identification in the event it is necessary to identify and segregate lots of food that may be contaminated.

- Records should be kept for a period of time that exceeds the self life of the product, except that
- Records need not be kept beyond two years.

Hazard Analysis and Critical Control Points (Haccp)

Hazard analysis and critical control points, or HACCP /ÈhæsŒp/ , is a systematic preventive approach to food safety and biological, chemical, and physical hazards in production processes that can cause the finished product to be unsafe, and designs measurements to reduce these risks to a safe level. In this manner, HACCP is referred as the prevention of hazards rather than finished product inspection. The HACCP system can be used at all stages of a food chain, from food production and preparation processes including packaging, distribution, etc. The Food and Drug Administration (FDA) and the United States Department of Agriculture (USDA) say that their mandatory HACCP programmes for juice and meat are an effective approach to food safety and protecting public health. Meat HACCP systems are regulated by the USDA, while seafood and juice are regulated by the FDA. The use of HACCP is currently voluntary in other food industries.

HACCP is believed to stem from a production process monitoring used during World War II because traditional "end of the pipe" testing on artillery shell's firing mechanisms could not be performed, and a large percent of the artillery shells made at the time were either duds or misfiring. HACCP itself was conceived in the 1960s when the US National Aeronautics and Space Administration (NASA) asked Pillsbury to design and manufacture the first foods for space flights. Since then, HACCP has been recognized internationally as a logical tool for adapting traditional inspection methods to a modern, science-based, food safety system. Based on risk-assessment, HACCP plans allow both industry and government to allocate their resources efficiently in establishing and auditing safe food production practices. In 1994, the organization of *International HACCP Alliance* was established initially for the US meat and poultry industries to assist

Certified HACCP Auditor (CHA) in 2004. HACCP expanded in all realms of the food industry, going into meat, poultry, seafood, dairy, and has spread now from the farm to the fork.

The HACCP Seven Principles

Principle 1: Conduct a hazard analysis. – Plans determine the food safety hazards and identify the preventive measures the plan can apply to control these hazards. A food safety hazard is any biological, chemical, or physical property that may cause a food to be unsafe for human consumption.

Principle 2: Identify critical control points. – A critical control point (CCP) is a point, step, or procedure in a food manufacturing process at which control can be applied and, as a result, a food safety hazard can be prevented, eliminated, or reduced to an acceptable level.

Principle 3: Establish critical limits for each critical control point. – A critical limit is the maximum or minimum value to which a physical, biological, or chemical hazard must be controlled at a critical control point to prevent, eliminate, or reduce to an acceptable level.

Principle 4: Establish critical control point monitoring requirements. – Monitoring activities are necessary to ensure that the process is under control at each critical control point. In the United States, the FSIS is requiring that each monitoring procedure and its frequency be listed in the HACCP plan.

Principle 5: Establish corrective actions. - These are actions to be taken when monitoring indicates a deviation from an established critical limit. The final rule requires a plant's HACCP plan to identify the corrective actions to be taken if a critical limit is not met. Corrective actions are intended to ensure that no product injurious to health or otherwise adulterated as a result of the deviation enters commerce.

Principle 6: Establish procedures for ensuring the HACCP system is working as intended. – Validation ensures that the plants do what they were designed to do; that is, they are successful in ensuring the production of a safe product. Plants will be required to validate their own HACCP plans. FSIS will not approve HACCP plans in advance, but will review them for conformance with the final rule.

Verification ensures the HACCP plan is adequate, that is, working as intended. Verification procedures may include such activities as review of HACCP plans, CCP records, critical limits and microbial sampling and analysis. FSIS is requiring that the HACCP plan include verification tasks to be performed by plant personnel. Verification tasks would also be performed by FSIS inspectors. Both FSIS and industry will undertake microbial testing as one of several verification activities.

Verification also includes 'validation' – the process of finding evidence for the accuracy of the HACCP system (e.g. scientific evidence for critical limitations).

Principle: Establish record keeping procedures. – The HACCP regulation requires that all plants maintain certain documents, including its hazard analysis and written HACCP plan, and records documenting the monitoring of critical control points, critical limits, verification activities, and the handling of processing deviations.

Standards

The seven HACCP principles are included in the international standard ISO 22000 FSMS 2005. This standard is a complete food safety and quality management system incorporating the elements of prerequisite programmes (GMP & SSOP), HACCP and the quality management system, which together form an organization's Total Quality Management system.

HACCP Training

Training for developing and implementing HACCP Food Safety management system are offered by several quality assurance companies. However, ASQ does provide a Trained HACCP Auditor (CHA) exam to individuals seeking professional training. In the UK the Chartered Institute of Environmental Health (CIEH) offers a HACCP for Food Manufacturing qualification. accredited by the QCA (Qualifications and Curriculum Authority).

HACCP Application

Applied Range: It can apply to several food categories; sea food, bulk milk production line, Bulk Cream and Butter Production Line, animal meat industry, Organic Chemical Contaminants in Food, Corn Curl Manufacturing Plant, etc.

USA

- Fish and fishery products
- Fresh-cut produce
- Juice and nectary products
- Food outlets
- Meat and poultry products
- School food and services

HACCP Implementation

It involves monitoring, verifying and validating of the daily work that is compliant with regulatory requirements in all stages all the time. The differences among those three types of work are given by Saskatchewan Agriculture and Food.

HACCP Versus ISO 22000

ISO 22000 is a standard designed to help augment HACCP on issues related to food safety. Although several companies, especially the big ones, have either implemented or are on the point of implementing ISO 22000, there are many others which are hesitant to adopt it. The main reason behind that is the lack of information and the fear that the new standard is too demanding in terms of bureaucratic work, from abstract of case study.

Bibliography

Adrian, Cullis, and Pacey, Arnold,: *Rainwater Harvesting*, Intermediate Technology Publications, UK, 1986.

Altieri, Miguel A. *Agroecology*: *The Science of Sustainable Agriculture*. Westview Press, Boulder, CO., 1995.

Arnold L. Aspelin. *Pesticides Industry Sales and Usage: 1994 and 1995 Market Estimates,* U.S. EPA, 1997.

Bard, A.J.; Faulkner, L.R.: *Electrochemical Methods: Fundamentals and Applications*, John Wiley & Sons, New York, 2005.

Brady, N.C., and Ra.R. Weil: *The Nature and Properties of Soils*, Prentice Hall, Upper Saddle River, NJ, 2002.

Bucks D.A., Nakayama F.S. and Gilbert R.G.: *Trickle Irrigation Water Quality and Preventive Maintenance*, Agricultural Water Management, NY, 1979.

Clark, Andy (ed.) Managing Cover Crops Profitably, *Sustainable Agriculture Network*, Beltsville, MD, 2007.

Conway, D.: *Climate Change and Water Resources in the Nile Basin*, University of London, London, 1993.

Dhaka , J.M.: *Economics of Agricultural Production and Farm Management*, Aavishkar Pub, Delhi, 2009.

Doneen L.D.: *Water Quality for Agriculture*, Department of Irrigation, University of California, Davis, 1964.

Douglas, J.S.; R. A. de J. Hart. *Forest farming: towards a solution to problems of world hunger and conservation*. Intermediate Technology Publications, London, 1984.

Elmusa, Sharif S.: *Water Conflict: Economics, Politics, Law and Palestinian-Israeli Water Resources*, Institute for Palestine Studies, Beirut, Lebanon, 1998.

Erik Nissen-Peterson: *Rainwater Catchment Systems*, UK: Intermediate Technology Publications, 1999.

Ertel, Patrick W.: *The American Tractor: A Century of Legendary Machines*, MBI, Osceola, WI, USA, 2001.

Frasier, Gary, and Lloyd Myers: *Handbook of Water Harvesting*, U.S. Dept. of Agriculture, Agricultural Research Service, Washington D.C, 1983

Goyal Sham S.: *Crop Production in Saline Environments: Global and Integrative Perspectives*, International Book, Delhi, 2004.

Guthman, J. *Agrarian Dreams: The Parodox of Organic Farming in California*, Berkeley and London: University of California Press, 2004.

Hanan, J. J.: *Greenhouses: Advanced Technology for Protected Horticulture.* Boca Raton, Fla.: CRC Press, 1998.

Hemenway, Toby: *Gaia's Garden: A Guide to Home-Scale Permaculture*, Chelsea Green Publishing Company, Vermont, 2000.

Jana B L : *Water Harvesting and Watershed Management*, Agrotech, Delhi, 2008.

Khan, M A: *Water Resources Management and Sustainable Agriculture*, APH, Delhi, 2008.

Klancher, Lee; Leffingwell, Randy; Morland, Andrew; Pripps, Robert N.: *Farm Tractors*, Crestline, 2003.

Lowes, P.: *Developing World Water*, Grosvenor Press International, London, 1987.

Malcolm D. Bolton: *A Guide to Soil Mechanics*, Universities Press, Delhi, 2003.

McGarry, M.G.: *Matching Water Supply Technology to the Needs and Resources of Developing Countries,* United Nations. New York. 1987

Pacey A., and A. Cullis: *Rainwater Harvesting: The Collection of Rainfall and Runoff in Rural Areas,* London, U.K.: IT Publications, 1986.

Prabakaran, G.: *Introduction to Soil and Agricultural Microbiology*, Himalaya, Delhi, 2004.

Samuel W. Johnson: *Crops Feed from Air and Soil*, Reprint Pub, Delhi, 2005.

Sanghvi Sheela and Pahwa Prem S.: *Water Harvesting, Purification and Distribution Management* , Dominant, Delhi, 2001.

Sharma, Premjit: *Applied Soil Ecology*, Gene Tech Books, Delhi, 2007.

Singh S.D.: *Water Harvesting in Desert*, Manak, Delhi, 1997.

Thapliyal, B.K.: *Democratisation of Water*, Serials Pub, Delhi, 2008.

Thomas, Harold E.: *The Conservation of Groundwater: A Survey of the Present Groundwater Situation in the U.S.,* McGraw Hill, New York, NY. 1951.

Wojtkowski, Paul A. *The Theory and Practice of Agroforestry Design.* Science Publishers Inc., Enfield, NH, 1998.

Wojtkowski, Paul A. *Agroecological Perspectives in Agronomy, Forestry and Agroforestry.* Science Publishers Inc., Enfield, NH, 2002.

Index

❑❑❑